The Cambridge Manuals of Science and
Literature

LIFE IN THE SEA

Benthos and Nekton.

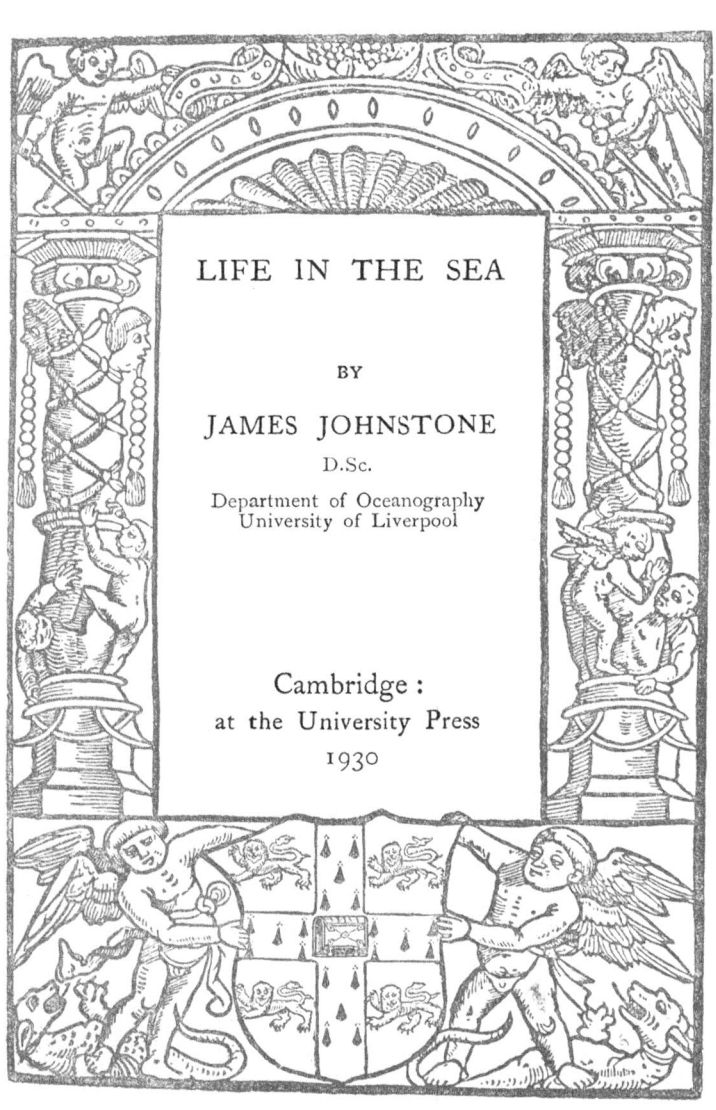

LIFE IN THE SEA

BY

JAMES JOHNSTONE

D.Sc.

Department of Oceanography
University of Liverpool

Cambridge :
at the University Press
1930

CAMBRIDGE UNIVERSITY PRESS
Cambridge, New York, Melbourne, Madrid, Cape Town,
Singapore, São Paulo, Delhi, Tokyo, Mexico City

Cambridge University Press
The Edinburgh Building, Cambridge CB2 8RU, UK

Published in the United States of America by Cambridge University Press, New York

www.cambridge.org
Information on this title: www.cambridge.org/9781107404120

First Edition 1911
Second Edition 1930
First paperback edition 2011

A catalogue record for this publication is available from the British Library

ISBN 978-0-521-05444-7 Hardback
ISBN 978-1-107-40412-0 Paperback

*With the exception of the coat of arms at
the foot, the design on the title page is a
reproduction of one used by the earliest known
Cambridge printer, John Siberch, 1521*

PREFACE

THE history of every science affords examples of the enormous stimulus to further research which arises from the discovery of some new and sound method of research ; and two such advances in methods have transformed marine biology during the last ten years. Late in the eighties of last century some German physiologists began to estimate the quantity of microscopic life contained in the sea under each square metre of surface, and nets and other apparatus were invented for this purpose. About the beginning of the nineties some Scandinavian geographers and zoologists also began to investigate the microscopic life of the sea with respect to the physical conditions of the water. The object of the Kiel quantitative plankton investigations was to estimate the production of a sea area in much the same way as a scientific agriculturalist attempts to estimate the productivity of a cultivated land area ; and that of the Scandinavian hydrographers was to attempt to forecast seasonal changes in the climatic conditions of their countries by studying the currents of the sea.

Now apart altogether from the primary objects of these researches results were obtained which have proved to be of great theoretical interest, so that the original methods employed have been improved, and their use has been extended. There is of course a very great amount of such work still to be done but we know enough even now to make it well worth while to attempt to make a broad survey of life in the ocean; to consider the way in which the different kinds of organisms affect each other; and how they are influenced by the great seasonal changes which sweep across the sea. Such a discussion of the general economy of the sea is attempted in this book.

It may seem to some readers who possess a first-hand acquaintance with some of the subjects dealt with, that many of the statements made are too dogmatic. It is impossible to give all the evidence that might be brought forward in support of them, and I have therefore added some notes at the end of the book, which indicate the main sources of information. Finally one may admit that there is difference of opinion with regard to some of the views stated, but a good deal of evidence may be adduced in their favour; and at all events they are permissible views in the present state of our knowledge.

JAS. JOHNSTONE.

Liverpool,
  September, 1911.

In the new edition, the section on the authorities has been brought up to date and re-set, and minor corrections and additions have been made to the text.

CONTENTS

Frontispiece. Benthos and Nekton; a coral
    bank in the Red Sea. (From Haeckel.)

The Tailpieces represent various pieces of
    oceanographic apparatus   pp. 27, 52, 86,
    117, 142

CHAPTER I

THE CATEGORIES OF LIFE

It is probable that there are at least about 500,000 distinct species of plants and animals in the sea. Each species is an assemblage of individuals which differ from the individuals composing other species or assemblages, in form, in life-history, and to some extent in habits. Above all, the individuals of each species differ from those of other species in that their mode of reproduction has been so specialised, that the offspring born from them resemble the parents more than they resemble the individuals of any other species. Now it is the task—as yet an incompletely performed one—of marine biology to classify all these organisms so as to display their inter-relationships to each other and to the species of organisms which have inhabited the earth in past geological periods. Even to indicate the main principles of such classification would be beyond the scope of this book, but if we assume a general knowledge of the main types of plant and animal life it will not be difficult to devise a grouping which will be at once convenient and philosophical.

Let us suppose that it is possible to start from some place on the west coast of Ireland and walk out along the sea-bottom, underneath the North Atlantic Ocean, until we reached the opposite shore of North America. As we did so we should witness a remarkable series of changes in the physical condition of the sea-bottom, and of the water over our heads ; and we should also witness a similar change in the nature of the life of the sea ; such a change as would enable us to associate the variations in the physical conditions with certain variations in the abundance and nature of the animal and plant life of the ocean. We will suppose that we were to start out from some stretch of beach covered by stones and gravel: on the higher parts of such a beach we should find a particular kind of life—large sea-weeds, such as the bladderwrack, clinging to the stones; mussels, barnacles, periwinkles, dog-whelks and small crustacea such as the sandhoppers. The animals and plants of which these are the most familiar examples inhabit the sea beach between the upper and lower tide marks and we may call them the Littoral Benthos. Now descending the beach to near the limit of low water of high spring tides we should find a more varied and luxuriant fauna and flora, which would consist of the large-fronded sea-weed Laminaria,

small fishes such as the blennies, gobies, and gunnells;
many crustacea such as lobsters, crabs, hermit-crabs,
and prawns; molluscs such as the whelks, horse-
mussels, clams, scallops, and brightly coloured sea-
snails (Nudibranchs); starfishes, sea-urchins and
brittle stars; many kinds of crawling worms; sea-
anemones, sponges, and numerous species of plant-like
zoophytes and polyzoa encrusting or growing on the
stones or larger algae. This fauna and flora inhabiting
the zone of shore only exposed by exceptional ebb-
tides we may call the Laminarian Benthos. If our
initial point of departure had been a rocky coast we
should still find much the same distinction, for the
upper parts of the rocks would be covered with the
barnacles, periwinkles, etc., while the lower rock-pools
would contain much the same organisms as we found
on the lower parts of the gravelly beach. If we had
started from a flat, sandy shore the nature of the
life would, however, be quite different, for the sand
between the tide marks would probably contain such
molluscs as the cockle, animals like the lugworm
burrowing beneath the surface, and small crustacea
such as the Copepods. The larger rooted algae would
be absent, but their place would be taken by microscopic plant life. Here and there on the surface of
the sand, on the ripple marks, or in the little gutters
left by the retreating tide there might be patches of
yellow-green slime, and on examining these we should

find them to consist of masses of Diatoms; and almost everywhere the sand would contain these though they might not be visible to the naked eye. The impression we should get would be that the sandy shore contained less life than that found on the stones or gravel, and this impression would probably be an accurate one: nevertheless the sand would contain far more life than one would at first imagine it to hold.

As we walked out to sea to greater depths we should witness many changes in the nature of the sea-bottom. Here and there it would consist of bare rock, or of stones and gravel, and the nature of this rock or gravel would depend on that of the adjacent land. But as we went further the more common deposit on the bottom would be sand, although there might be patches of mud, particularly in the deeper parts or channels. The general arrangement would be first of all rock and stones or gravels, then sands, and then muds; for all these materials result from the erosion of the land, and the gravel is first deposited or rolled down, then the sand, and lastly the mud. Out beyond a depth of about 50 fathoms the muddy deposits would be more abundant than closer inshore, for the finer particles of which they are composed are carried for a greater distance in the water, while much of the fine particles carried in suspension by the water of the rivers is precipitated by the salts of the sea-water and only settles down very slowly.

This mud would form in the deeper depressions where the run of tide is weakest, and we should never find it on ridges of rock standing up from the sea bottom. Now all these materials come from the wasting away of the land, but we should also find sea-bottom deposits which are formed by the agencies of living creatures. There would be patches of hard, sandy substance formed by the tubes of the worm Sabellaria; here and there deposits of the coral-like skeletons of calcareous algae, or Corallines; and perhaps also extensive deposits of broken shell fragments. So far we should have got down to a depth of perhaps 50 fathoms.

The life inhabiting the sea-bottom would not differ much from that which we had already observed in the Laminarian zone, but there would of course be greater variety, especially in regard to the fishes. The latter would consist of plaice, dabs, soles, gurnards, skate and ray, codling, whiting, dogfishes, and others. There would be shrimps on the shallower sandy bottoms, accompanied by hosts of invertebrates such as various species of crabs, starfishes, ophiurids, sea-urchins, hermit-crabs, polyzoa and many others. In the deeper places near the 20-fathom line there would be other species of fishes not found in the shallower water, such as turbot and brill, and there would also be a difference in the invertebrates : the larger squids and cuttle-fishes and many others would

be found here. The rocky bottoms would harbour
many species of fishes not found on the sand—the
brilliantly-coloured wrasses for instance, with perhaps
pollack, ling or saithe. The muddy places would
contain many kinds of burrowing crustacea and
worms, and perhaps the Norway lobster (Nephrops).
All this part of the sea we might call the shallow
water region, and the bottom life inhabiting it the
Shallow Water Benthos. Life here is richer and
more varied than in any other part of the sea, not
only because of its permanent fauna but also because
hosts of young fishes inhabit this region for the first
year or so of their lifetime. As we proceed further
out to sea there would be a progressive change both
in the nature and the abundance of life. The algae
would disappear entirely. The shallow water fishes
such as the plaice and sole would be replaced by the
witch and megrim, and the cod, whiting and haddock
by the hake. Invertebrate life would also become
rather less abundant. Between the 50-fathom and
the 100-fathom contour lines the region of sea may
be called the deep-sea area and its bottom life the
Deep-Sea Benthos. It would all be familiar to us.

So far we have been travelling under relatively
shallow water, over the narrow selvage of sea-bottom
which fringes the margins of the continents, and
extends out to sea for a variable distance from the
dry land. The British Islands themselves are situated

on this margin, which is called the 'Continental Shelf.' The slope downwards would be a fairly gradual one so far, but at depths of about perhaps 500 fathoms it would begin to become much more steep, and in some parts of the ocean it would be represented by a series of terraced precipices. Beyond these, and at depths of over 1000 fathoms, would be the permanent abysses of the ocean.

Very strange and unlike anything on the dry land would be this ocean bottom. Beyond the edge of the continental shelf there would be few hills, or valleys, and the slopes downward to greater depths, or upwards to oceanic islands, would be so gradual as to be hardly perceptible to us. Sailors, visualising the sea-bottom from their knowledge of it acquired by soundings, speak of knolls, deeps, etc., but they usually exaggerate the degree of slope actually existing. It is said that there are some fissures on the ocean bottom which resemble the deep river gorges of the land, or bare pinnacles and ridges of primeval rock uncovered by any deposits, but these must be quite exceptional features, and the configuration of the ocean floor probably resembles nothing so much as that of the great prairie lands of the western continents. But nowhere on the land would there be such immense tracts of level surface as those of the beds of the Atlantic, and the greater part of the Pacific, Oceans.

The materials forming the sea floor at shallow depths would mostly be familiar to us, but near the edge of the continental shelf they would become quite unlike anything that we know on the land. All the coarser detritus washed away from the land surface would have settled down, and everywhere the sea-bottom would be covered with soft, fine, muds or oozes, varying in colour through tints of blue, green, red or grey. These are the terrigenous muds and they consist of the very finest particles of mud borne in suspension by rivers and currents, and floating in the water for immense distances. Nearly everywhere they would be modified in composition by the action of decomposing organic matter resulting from the dead bodies of marine organisms. Here and there would be volcanic debris, scoriae, fine dust, or even fragments of lava, either ejected from submarine volcanic vents, or settling down on the ocean floor after having been drifted about in the sea or in the atmosphere. Beyond this area of terrigenous muds we should encounter that of the pelagic deposits ; and nearly everywhere at depths of 1500 fathoms to rather over 3000 fathoms the sea bottom deposits consist mainly of the shells or skeletons of animals and plants which live in the upper layers of the sea, and dying sink down to the bottom to form the abyssal oozes.

In many parts of the world the characteristic

deposit at depths of about 500 to 1500 fathoms is that known as Pteropod ooze. This consists to a large extent of the shells of the pelagic mollusca known as Pteropoda. We should not encounter any patches of this ooze in our traverse across the North

Diatom ooze   Pteropod ooze

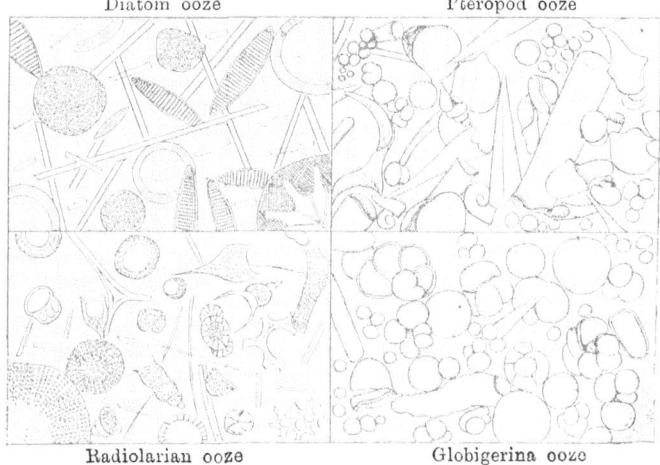

Radiolarian ooze   Globigerina ooze

Fig. 1. Deep-Sea Deposits.

Atlantic where the greater part of the sea-bottom, outside the limit of distribution of the terrigenous muds, consists of Globigerina ooze. This is a soft white or grey mud containing gritty particles and composed largely of carbonate of lime. The picture

on page 9 represents it after the finer amorphous
sediment which makes up the greater part of its
bulk is washed away.   The rounded shells of which
it is composed are those of pelagic Foraminifera,
protozoa which live mainly in the upper parts of
the sea, and which dying there sink down to the
bottom, where their shells accumulate to form the
deposit.   How deep it is we have no opportunity of
knowing, but telegraph engineers believe that it forms
at the rate of about an inch in ten years in some
places.   It is the characteristic deposit of the North
Atlantic ocean floor, and only here and there the
deeper parts of the latter are covered with material
of volcanic origin.   At depths over 2500 fathoms the
Globigerina ooze tends to disappear, for the lime
becomes dissolved by the sea water at these depths.
The deposit where the sea is about 3000 fathoms in
depth elsewhere than in the Atlantic consists of
either volcanic material or the mud called Radiolarian
ooze.   This latter formation consists largely of the
skeletons of the pelagic protozoa known as Radiolaria.
It contains very little carbonate of lime, for the
skeletons of the Radiolaria are composed of silica.
We should also find the skeletons of these organisms
in the deposits in shallower parts of the ocean, but
they are not so evident because of the greater mass
of the limy shells of the Foraminifera.   The siliceous
skeleton of the Radiolarians is better able to withstand

the solvent action of the sea water. This ooze is not the only one which is composed of siliceous material, for in the Antarctic there is an area of ten and a half millions of square miles covered with the remains of diatoms. These organisms live in the superficial layers of the sea in such enormous numbers that their dead shells form the major part of the mud covering the floor of the Antarctic.

In the greatest depths of all, 3500 fathoms and over, we find no longer any obvious indications of organic structures in the mud on the sea-bottom. The deposit in these great depths is called the Red Clay, and it is a very soft brown mud consisting of exceedingly fine particles, which are the insoluble remains of the shells of pelagic plants and animals, and volcanic dust settling down from suspension in the water, or from the atmosphere. Among it we find cosmic dust—that which has originated in the combustion of meteorites entering the atmosphere of the earth from outer space. It accumulates with exceeding slowness, for we ought to find meteoritic dust in every part of the sea-bottom, but it can usually be recognised only among the red clay, for everywhere else it is hidden by the more abundant materials. In the red clay we also find the earbones of whales and the teeth of sharks. All the rest of the skeletons of these great creatures have been dissolved by the action of the sea water.

Nothing in our ordinary experience gives us any adequate idea of the physical conditions at the bottom of the ocean. As we descend deeper and deeper the temperature falls more and more, and at a depth of about 400 fathoms the annual variations in the temperature of the sea due to the seasons fade away, and uniformity reigns. From about 400 fathoms downwards the temperature would fall very slowly, and everywhere in the North Atlantic it would not be much higher than that of the freezing point of fresh water. In the circumpolar seas it would fall still further and in some places would be about two degrees below zero of the Centigrade scale.

The darkness would be greater than that of which we have any ordinary experience. Even at a depth of about ten fathoms the illumination would be that of faint twilight only, and at fifty fathoms it would be still less. The light there would be no longer white, for sea water absorbs first the red, orange, and yellow rays, and then the green, blue and violet. At about 50 fathoms the violet light would still be fairly strong, and it would be present even at a depth of 200 fathoms, but there would be absolutely no trace of the red rays. At about 400 to 500 fathoms even the violet light would have disappeared entirely and only the ultra-violet rays—light which is invisible to our eyes, would remain. At a depth of about 900 fathoms the last traces of ultra-violet radiation would have disappeared and a delicate photographic plate exposed here for over two hours would not be affected. Absolute darkness exceeding that of the photographer's dark room would prevail.

We can easily imagine the cold, and the darkness, and the great plains of semi-liquid mud, but not the extreme pressure of the water. At the surface of the sea an organism is exposed to a pressure of one atmosphere, that is 15 pounds per square inch of its surface, and a rise of even one atmosphere produces very unpleasant effects in ourselves. But for every ten metres (5 fathoms) that we descend, the pressure due to the weight of the water rises by one atmosphere, and at a depth of 3000 fathoms the pressure on every square inch of the body of an animal amounts to about three tons. If we did not know that no part of the sea-bottom—however deep—is utterly devoid of animal life we might conclude—as the older naturalists did—that this enormous pressure was inconsistent with a belief in the existence of living organisms. Even now it is difficult to understand how the tissues of an animal are able to withstand it. As it is, the pressure exerts a profound influence on the structure of the animals inhabiting even such moderate depths as 200 fathoms. A deep-sea fish is easily recognised by its large eyes, its long and attenuated tail, and the softness of its flesh. One's finger leaves a dent in the flesh of such a fish as a hake if pressed against it. But the pressures in the abyssal regions are very much greater than those which a hake has to withstand. Even the thermometers at first used in deep-sea exploration were frequently shattered to fine dust by the pressure of the water ; and this is the great difficulty in the construction of all instruments for the investigation of the oceanic depths.

Just as we recognise a littoral, a shallow water and a deep-sea Benthos, so we may speak of an Abyssal Benthos inhabiting the whole sea-bottom deeper than 2000 fathoms. Before the time of the great oceanographic expeditions there was much speculation as to the nature of the animals which might be found in the greatest depths of the sea. It was thought that exploration might disclose unfamiliar and bizarre forms of life, perhaps gigantic animals, perhaps remnants of former geological faunas persisting under the uniform conditions of the abysses, while all the rest of the earth was undergoing physical change. These expectations have not been realised, for all the animals captured from the deep belong to well-known groups of life ; and with the exception of some stalked Crinoids, animals closely allied to the Crinoids which formed so characteristic a feature of the fauna of the Carboniferous period, most of the species belong to recently evolved groups. One cannot say that there are no monstrous or gigantic animals still living in the abysses, for no form of fishing gear yet used would enable us to capture a very large abyssal animal, and one may perhaps still attach some consideration to stories of sea-serpents. Nevertheless the general experience is that the animals obtained from the great depths of the sea are as a rule smaller than their shallow water allies. There is, of course, much that is strange in the abyssal Benthos, but though a zoologist accustomed only to shallow water animals would fail to recognise most of the deep-sea forms, he would still be able to refer them to their respective families or classes.

Deep-sea animals have certain general characters. Most of the great invertebrate groups, the molluscs, crustacea, sponges, echinoderms, alcyonarians, etc., are represented in the deep. The crustacea are often long-legged creatures coloured a uniform red. Most of them are smaller, or at least no larger than are the shallow water forms, but some are relatively gigantic—one of the Isopods, for instance, a group represented by the little 'slaters,' is about nine inches in length. Some of the crinoids, tunicates and alcyonarians (seapens) are carried on long stalks —apparently an adaptation to life in the soft oozes. There are echinoderms, starfishes and sea-urchins which do not differ in any essential respect from those living in shallow water except that spines are rather more pronounced in the abyssal forms.

The fishes are very unfamiliar to us though they all belong to well-known groups, and it is very probable that they have been evolved from shallow water species. They have as a rule either very large eyes, or very small ones, or they are quite blind. This is what we might expect if the fishes had spread into the deep from the shallow water, for evolution might proceed in various directions: either an organ which was inefficient under the new conditions would become larger by the accumulation of small variations, or being unused and useless it would disappear. In most of them the tail is long and slender; the mouth and teeth are very large so that the animal has an eminently predatory character, which would be a consequence of the difficulty in procuring food; some abyssal fish are able to devour others almost as large as themselves. Their bones are fragile and have less than the usual proportion of lime; a character which is common to most abyssal animals which have calcareous external or internal skeletons. The flesh is soft and watery. There is a general absence of the pronounced colour markings of the shallow forms; they are not silvery as a rule; and there are no indications of the obliterative counter-shading of the fishes usually known to us whereby the upper parts are darker than the lower ones so as to destroy the appearance of solidity, and render the animal nearly invisible to another in the water at the same level as itself. The colour is usually a monochrome—a black or dark red: in the deep, in the absence of red light, such a coloured fish would appear black; if indeed one can speak at all of colour in such surroundings.

The great majority of abyssal animals are phosphorescent: they possess luminous organs similar to eyes in structure, and often the fishes of moderately deep water have rows of such ocelli along their sides. Sometimes deep-sea animals secrete a sort of phosphorescent slime. Why they should produce light it is hard to guess: we must remember that it is only to our eyes that they appear luminous. Very strange indeed would be the appearance of these animals if we could see them in the deep. In the absolute darkness of the abyss they would appear as ghostly silver-blue shapes glimmering like an electric lamp through dense fog on a dark moonless night. Of all the characters of the deep-sea fauna this almost universal phosphorescence is the strangest.

All the animals and plants which we have so far examined belong to the category of life called the Benthos—a collective term applied to those organisms living at the sea-bottom attached to stones or other fixed objects; or burrowing in the sand or mud; and which may also be extended so as to include all those animals which, though free to move about, are sluggish and inert, and which frequent only a

comparatively small area of sea-bottom. Among the
attached forms we find the algae, the zoophytes and
polyzoa, the sponges, the reef-building and solitary
corals, the sea-anemones, the branching corals and
alcyonarians, the crinoids, some crustacea of which
the barnacle is the most familiar example, many
molluscs such as the cockle, mussel and oyster and
most of the tunicates. Among the burrowing forms
are some protozoa, many worms, some sea-urchins,
and sea-cucumbers (Holothurians), some sea-anemones,
many crustacea, and many molluscs such as the clams
or razor-shells. The sluggish, slowly moving, bottom
living animals include the starfishes and most of the
sea-urchins, crabs, lobsters and hosts of other crustacea,
many worms, many molluscs such as the scallops,
periwinkles and whelks. Some fishes are also to be
placed in this category. Benthic animals thus include
representatives of all the great groups of life. When
we remember that distinct faunas are to be found
on sea-bottoms of different depths, and of different
physical conditions, we see that it is convenient to
speak of littoral, shallow water, deep water, and
abyssal benthic populations. In the wider economy
of the sea all these categories of life are distinct from
each other.

Now the great majority of marine animals are
benthic in habit but there are a number which possess
powerful locomotory organs and which are thus able

to migrate over wide areas of sea. These are all grouped together to form the category of life known as the Nekton. Comparatively few marine animals in addition to the fishes are nektic in habit. The whales, seals, dolphins and porpoises are the next largest group, and in addition to these and the fishes only a few invertebrates, such as the squids, the cuttle-fishes, and the giant calamaries are actively moving animals; and perhaps some pelagic worms and medusae may also so be described. According to their habits the fishes may be further subdivided into demersal and pelagic species. The demersal fishes are those which live on the sea bottom and obtain their food from animals which also live there. Most of the commoner fishes belong to this category; the sole, plaice, whiting, haddock, cod, skate and ray, turbot and brill and many others are examples. The pelagic forms are the minority, and the best-known examples are the herring, mackerel, sprat, pilchard, and probably also the salmon while it inhabits the sea.

As a rule the distinction between demersal and pelagic fishes is a very real one, and it is exceptional for a fish which lives at the sea bottom to approach the surface; if it does the habit may be regarded as an abnormal one. If we trawl demersal fish such as the cod or whiting from even the moderate depth of 20 fathoms, and take care that the animals are not injured by the action of the fishing gear, we see

2—2

that they have the greatest difficulty in again going down to the sea-bottom, and for a time they struggle helplessly at the surface belly upwards. This is because of the expansion of the gas contained in their swim-bladders, and it is only after this has been absorbed by the blood stream that the fish regains its former specific gravity and is able to go below the surface. It used to be the habit of the Grimsby trawlers, who often brought cod alive into the markets, to prick the fish behind the shoulder with a long needle and so allow the gas in the swim bladder to escape. On the other hand fish like the plaice and sole, which have no swim bladders, are able to be transferred directly from water of moderate depths to quite shallow tanks without any apparent bodily disturbance, even although the difference in pressure may be more than one atmosphere. Pelagic fish are able to live both at the surface of the sea and in the depths—say of ten to twenty fathoms; thus the herring must go down to the sea-bottom in order to spawn, and they are often caught by means of the trawl-net in the same manner as the proper demersal fishes are captured. But we may be sure that these fishes are able by means of some control over the blood vessels of the swim bladder to alter their specific gravity.

In the case of the fishes of moderately shallow water—say down to twenty fathoms—there is little, if any essential difference in their form: a herring, for instance, does not differ greatly in structure from a whiting, though the one is a pelagic, surface-living animal and the other is one which nearly always lives at the sea-bottom, and both have swim bladders. But in the case of the oceanic fishes the structure may be strikingly different according to the depth of water which the animal usually inhabits. A flying fish differs very notably from any of the abyssal forms which we have already considered; and those species which live at intermediate depths are also very remarkable in form. They are, as a rule, small, very delicate in structure, with semi-transparent bodies, often compressed from side to side. It is quite easy to read ordinary print through the body of the peculiar Leptocephalus larva of the eel. Many of them are tinted in shades of violet or silver-blue, and have a peculiar shimmering appearance. Often their eyes telescope into sockets. As a rule these are the characters of the pelagic fishes found in the oceans, in the intermediate layers down to a depth of about 200 fathoms. Below that level the colour is usually black or dark red, and the form of the fish is that which we have already recognised as characteristic of the abyssal category.

All the animals and plants which we have dealt with so far are such as can easily be seen with the naked eye, or which can be captured by means of the nets and lines of the fisherman. But everywhere the sea water, no matter how clear it may be, contains abundance of life in the form of organisms too small to be seen without the aid of the microscope. We can always capture these organisms by filtering even a tumblerful of water through a sieve made of the very fine silk cloth that millers use in separating the various grades of flour. It would be rare indeed for as much sea water as could be lifted in an ordinary thimble to be without a dozen or more microscopic organisms, and at the time of maximum abundance of life in the sea such a quantity of water might contain several thousands of living creatures. All these organisms which can so be obtained are grouped together to form our third great category of marine life—the Plankton. The general character of all is that they are unable to make migrations which can materially influence their distribution. Many of them do not possess any organs of locomotion, and even if they do these organs are quite ineffective for that purpose. Planktonic animals and plants are simply drifted about passively in the sea by the agencies of tidal streams, currents, and winds.

The composition of the plankton is remarkably variable. Looking at it from a general point of view we can recognise in its constitution two great classes of organisms. (1) Larval stages of all the great groups of life: thus fishes which live either demersally or pelagically spawn eggs which, as a rule, drift about

in the upper levels of the sea-water. Some fishes
indeed, are viviparous—thus some of the dogfishes,
rays, and sharks produce their young alive, and some
others lay large eggs which incubate while lying at
the sea-bottom in sheltered places. But the majority

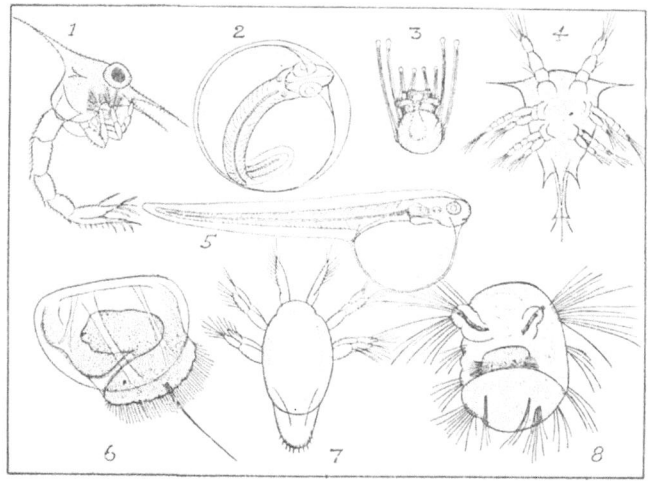

Fig. 2.   Plankton: Larvae.   1, Crab; 2, Fish-egg; 3, Sea-urchin;
   4, Barnacle; 5, Fish; 6, Mussel; 7, Copepod; 8, Worm.   All
   magnified.

of fishes produce young which live in the plankton as
microscopic drifting animals, for a period of their lives.
Many of the crustacea also breed in this way, and their
larvae are planktonic; so are most of the very young

mollusca, many of the starfishes and their allies, and
indeed most of the groups of invertebrata. All these
larval creatures form what we may call the transitory
plankton. They live drifting about in the sea for a
variable period of their lives, and then they settle

Fig. 3. Plankton. 1 and 2, pelagic worms; 3, a Medusa; 4, a
Copepod; 5, a Ctenophore. All magnified.

down to their final habitat as members of either the
Benthos or the Nekton. (2) The second and larger
part of the plankton consists of organisms which live
there permanently.

These latter creatures are far more important
from the point of view of the economy of the sea
than are the larvae. They constitute probably the

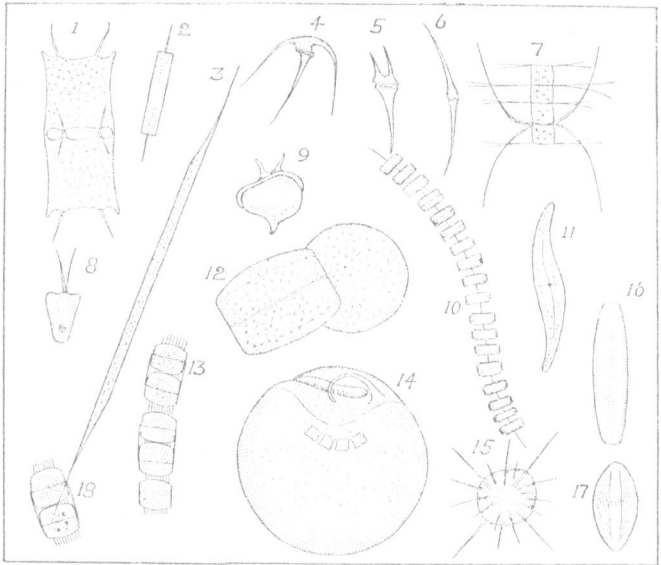

Fig. 4. Plankton : Unicellular organisms. 1, 2, 3, 7, 10, 11, 12, 13,
16, 17, 18, Diatoms; 4, 5, 6, 9, Peridinians; 8, an Algal reproductive Cell, or Zoospore; 14, Noctiluca; 15, a Radiolarian. All
magnified.

largest of all the categories of life in the sea, for
though they are very minute in point of size, yet

they are present in extraordinary numbers. It is quite probable that if we could suddenly dry up, say, one square mile of sea, we should find that the benthic and nektic animals and plants were really less in bulk than were the invisible organisms of the plankton. Now two great sub-kingdoms of life—the Algae among the plants, and the Protozoa among the animals—form the major part of the permanent plankton. The Algae are represented principally by the Diatoms, minute unicellular plants, possessing an outer skeleton of silica. Generally speaking the Diatoms constitute the bulk of the plankton during the spring, and they are nearly always present at any time of the year. Many sub-groups of Protozoa are always present, among them we may mention the Flagellates, Foraminifera and Radiolaria. These latter groups, with the Diatoms, are responsible for the formation of the abyssal sea-bottom deposits. Among the Flagellates one group—the Peridinians— is notable. It contains unicellular organisms which we may regard either as animals or plants, for they have the structure of the animal unicellular organism, but they have the same method of nutrition as the unicellular plant. They are creatures of fantastic shapes, with outer skeletons of a substance resembling excessively thin paper impregnated with silica ; and possessing chlorophyll, the colouring matter of the green plants.

Generally speaking the bulk of the plankton is made up of larval invertebrates, Diatoms, Peridineans, and some groups of micro-crustacea, the Copepods mainly, which live permanently in the plankton. There is this general relation between the abundance of the Diatoms, or Peridinians, and that of the Microcrustacea, such as the Copepods, that when one is abundant, the other group is scarce. There are, of course many other planktonic animals and plants. The larger rooted algae are represented by their zoospores. Many molluscs are permanently planktonic—thus the Pteropods are abundant enough to form the food of the whale, and to constitute a large part of the sea-bottom deposits in some regions of ocean. Many medusae and other jelly fishes are also planktonic.

The Naturalist's Dredge.

CHAPTER II

RHYTHMICAL CHANGE IN THE SEA

THE ocean is regarded as the very type of mutability and change, and indeed ordinary observation shows us that its condition is never quite the same from day to day. Many of the changes that occur in it—storms and calms, fogs, exceptionally high tides and the like, appear to us to be without order or regularity, that is their causes are so many and so complex that we cannot predict their occurrence. On the other hand there are many changes which are repeated so regularly as to constitute definite rhythms of events. Experience shows that their times can be calculated and that they recur over and over again with the certainty of astronomical phenomena. Such are the tides, the annual waves of temperature, salinity and sunlight; annual outbursts of animal and vegetable life; animal and plant migrations; spawning periods; fishery seasons and the like. Of all these periodic changes in the condition of the sea the purely physical ones are those which are repeated with the greatest regularity. Organic changes depend on these physical ones, but the inter-relationship is so complex that it is sometimes difficult to detect the periodicity.

About twice in every twenty-four hours the tide encroaches on the land, converting what a few hours ago were desolate expanses of sand and mud into evanescent sheets of water ; and again laying them bare to the atmosphere. Observing the order of the tides a little more closely we notice that they do not recur at precisely the same time, nor do they rise to exactly the same height on two successive days. Shortly after the time of full moon the water rises to its maximum, and then for about a week the flood tide culminates a little later every day—and rises a little less high. About halfway between the times of full moon and new moon the velocity of the tidal stream is least, and so also is the height to which the water rises. Then the tides begin to gather force and shortly after the date of new moon they again attain a maximum. Thus there is a half-daily rhythm, for a rise and fall of the tides occurs twice during each twenty-four hours ; but there is also a fortnightly rhythm, for spring tides and neap tides occur twice in each lunar month. If we continue to observe these things throughout an entire year we find also that every six months, about the times of the vernal and autumnal equinoxes, the spring tides rise to an unusual height. So there is a semi-annual tidal period, and on this is superposed a semi-lunar one, and on this again a semi-diurnal period.

Observation has shown that these rhythms are almost absolutely regular. Variations do indeed occur, for exceptionally strong winds blowing from the land prevent the flood tide from rising so high as it otherwise would, and cause the ebb tide to be lower than is predicted ; while strong winds blowing straight in from the sea have precisely the opposite effect. If the barometer rises one twentieth of an inch the flood tide will be an inch less than it would be if the atmospheric pressure were constant ; and if the barometer falls the reverse effect is produced. But these are only apparent irregularities, for with our present knowledge we cannot exactly foresee the variations in the pressure of the atmosphere, and apart from them almost everything is calculable or predictable. It is known for two or three years in advance, and for a great number of places all over the world, precisely how high the tide will rise, and when it will be high water every day throughout the year ; and so accurate are these tide-tables that sailors have come to accept them almost as part of the order of nature, thinking little of the wonderful mathematical analysis that has made them possible.

Ordinary observation shows again that once a

year a wave of temperature change sweeps across the sea. Sometime in our latitudes about February or March the sea is coldest and then, day by day, the temperature rises until it attains a maximum in August, and then it falls until the time of the next minimum. Just as surely as the earth revolves round the sun does the sea experience this annual temperature change; but if we study the variation day by day with a delicate thermometer we find that the rhythm is not a simple one. There are, as in the case of the tides, irregularities which are due to storms and other unpredictable occurrences, but there is also a daily rhythm. Just about sunrise the sea is coldest and then it gradually rises in temperature until about 4 p.m., the time of the daily maximum. Also there are rises and falls of temperature synchronous with the tidal periods: a few days after the time of full moon, or new moon, the sea near the land is warmer during the summer than it would be if there were no tidal streams; and during the winter it is a little colder. Thus there is a daily temperature period superposed upon a fortnightly one; and this again is superposed upon an annual period.

There are corresponding variations in the intensity of the sunlight that falls upon the surface of the sea. At about the end of the winter solstice this is at a minimum, for the sun's height is then lowest and its rays fall obliquely upon the surface of the sea. But day by day the sun rises earlier and sets later, and it also rises to a greater height, until the date of the summer solstice when the amount of light radiated on to the surface of the sea reaches a maximum. The intensity of sunlight passes through the same cycle of changes on the sea as on the land, and the times of maximum and minimum are the same in either case ; but this is not the case with the temperature change, for the minimum and maximum occur a little later in the sea than on the land. There is also a fortnightly period of light intensity in the sea, for once a month the moon passes through all her phases, and for a certain number of days reflected sunlight falls on the surface of the sea during the night. The amount of energy contained in the moonlight is not very great, but we shall see that only a very small proportion of the direct rays of the sun are used in the absorption of energy by plants, and it is probably the case that the light of the moon is strong enough to affect the rate of growth of planktonic organisms. Thus in the case of the intensity of sunlight there are daily, fortnightly and annual periods.

Even if we depended on ordinary means of observation, such as tasting the water, or observing the height of the waterlines of ships, we should be able to see that the salinity of the sea varied in different parts of the ocean of the earth : it is saltest in the Red Sea and freshest in some parts of the Baltic. But if we used delicate hydrometers, or still better chemical methods of estimating the proportion of salts, we should find that the salinity varies at the same place throughout the year. Now there are many irregularities in the changes of salinity which we cannot predict ; heavy rainfalls on the surface of the sea, or floods in the rivers opening into it, will produce local changes of salinity by diluting the sea-water. Nevertheless there are regularities as well. In some places we can detect a fortnightly variation which is due to the tides, where these run to and from the land. Thus, opposite the mouth of an estuary the sea will be, on the average, fresher some time shortly after the end of the ebb tide, for the water coming down from the river will have been slightly fresher than that which would have been present in the sea had there been no tidal streams, and this effect will be stronger at the period of spring tides, when the streams run with greater force.

There are also annual changes in the salinity, and to understand these we must consider the general scheme of oceanic circulation. The intense solar radiation in the tropics heats the surface of the sea, and the water becomes warmer and expands so that it stands at a higher level. It therefore flows

away to the north and south towards the poles, but
at about latitude 60° N. it has become cooled by
radiation of its heat to the atmosphere, and being
salter than the water normally present in the sea
at these latitudes it becomes heavier and sinks
towards the bottom.　Part of the water thus sinking
flows back again towards the equator.　Round the
poles the sea contains much ice, but when sea water
freezes most of the salt is squeezed out, and therefore when this ice melts it forms nearly fresh water.
The water of the circumpolar seas is therefore fresher
than normal sea-water and it accumulates at the
surface and then begins to flow towards the equator.
But it soon mixes with the other sea-water and
becomes much salter, and then being cold and heavy
it tends to sink to the bottom.　This occurs at about
latitude 60° N. (considering the Atlantic alone) and
part of this sinking water flows along the sea-bottom
back again to the poles.　Thus we have two main
systems of ocean currents, (1) warm and very salt
water flowing to the north and south from the
equator, and (2) cold and fresher water flowing also
to the south and north from the poles.　Now these
drifts of water do not actually flow to the north and
south : they would do so if the earth were entirely
covered with sea and if it did not rotate; but because
of the presence of the land the currents are diverted
in various ways ; and because of the rotation of the

earth they are continually deflected to the east in the northern hemisphere, and to the west in the southern hemisphere. In the North Atlantic the northerly flowing current is what we call the Gulf Stream; and the southerly flowing one consists of the East Greenlandic and Icelandic currents, and the Labrador current.

The Gulf Stream spreads out in the Atlantic in the form of a gigantic whirlpool the northern margin of which is never higher than the latitude of the Azores. Looking at an astronomical diagram we see that for six months in the year most heat falls on the part of the sea immediately north of the equator, while for other six months most heat falls on the sea immediately south of the equator. Therefore the equatorial stream shifts a little way north and south of the equator in the course of the year, and the whole area covered by the Gulf Stream circulation (which is caused by the equatorial stream) also shifts a little way north or south as the earth swings round the sun in the course of the year. There is just the same annual variation in the amount of water flowing south from the north pole, for more ice is melted during the summer than during the winter and therefore more fresh water forms and flows away to the south.

What we call the Gulf Stream in our latitudes is really the drift of comparatively warm and salt water proceeding to the north-east from the Gulf Stream eddy in the subtropical Atlantic. When the eddy has expanded farthest to the north the drift is strongest, and when it has retracted furthest to the south the drift is weakest. Now the water which comes up to the north-east, to the British Islands and coasts of Scandinavia, is always warmer and salter than the water in the North Atlantic would be normally if there were no Gulf Stream drift. Thus there is an annual change of salinity in our seas which depends mainly upon the strength of the Gulf Stream flow. The water is saltest when the drift is strongest, in the months of February to June, and it is less salt when the drift is weakest, in the months of November to February. The further north we go the later in the year does the Gulf Stream drift culminate. The amount of change that can be detected in the salt-ness of the sea and which is due to these causes is not much—the limits are about 34·5 to 35·5 per thousand. Salinity is defined in this way—as the total amount of salts weighed in grams and present in 1000 cubic centimetres of sea-water at a tempera-ture of 4° C.

Thus there is an annual period in the change of salinity of the sea and superposed on this are smaller periods which are due to various causes. But there are also longer periods than the annual one. There is a two-yearly period, for the sea is salter in the

springs of the 'even' years than it is in the springs
of the 'odd' years. There is also a period of about
twelve years and we have reasons for believing
that this is to be associated with the twelve-yearly
period of sunspot change. The years when there are
the greatest numbers of sunspots are also the years
when the solar radiation is most intense, for the
sunspots are the indication of increased activity in
the sun. Both the temperature and salinity of the
sea are subject to these long-period changes.

All these rhythmic changes, the tides, sea temperature and salinity, and intensity of sunlight are
of cosmic origin, and we can trace them back to
certain fundamental causes—the rotation and revolu-
tion of the sun, and the intensity of solar radiation.
In the case of the tides we can predict the rhythmic
change with great accuracy ; we can also predict the
annual temperature and salinity waves, but not the
exact dates of the maxima and minima ; and we can
predict with accuracy the times of sunrise and sunset,
and the daily declination of the sun, but not the
amount of cloud-covered sky. We know that the
sun is a variable star, but we do not know the exact
period of its variability, nor the range of variation
of its radiation.

All organic changes, that is, the periodic changes
in the habits of marine organisms, must ultimately
depend on these cosmic changes, but the inter-

relationships of the two series of events are so complex
that the organic changes are far less susceptible of
prediction than the physical ones. Nevertheless we
can easily show that very many changes in the nature
and abundance of life in the sea, and in the habits of
animals recur again and again with great regularity.
Let us take the case of the influence of the tides.
Nothing impresses one so much as the way in which
a fisherman regulates his practice according to the
tidal rhythms, so much so that he will associate the
latter with all sorts of variable occurrences with which
they can have nothing to do. The dependence of
littoral animals on the ebb and flow of the tides is
absolute, for it is only when the foreshore is covered
with water that they can feed or move about. Many
animals such as the cockle and mussel close their
shells when the tide ebbs off from the foreshore, and
others, such as the lugworm, burrow into the sand.
Convoluta, which is a little flat-worm living on the
sands in some parts of the coasts of France, also
burrows beneath the surface when the water leaves
it. These habits are easily explained, for a shell-fish,
such as a mussel, would become dried up if it were
to keep its shell open during the whole period of
ebb tide on a hot summer day, and the worms would
also become parched up by the heat of the sun. So
also with the spawning of Convoluta, which takes
place only once a fortnight at a certain phase of the

spring tide: the eggs are laid then because for a longer period than usual the sands are covered with water owing to the higher rise of the tide. But we find that this tidal rhythm is very deeply impressed on the conscious or unconscious memory of the animals. If we attempt to keep cockles in an aquarium for a long time it becomes difficult to keep the animals in good health, and we can improve the conditions by running away the water of the tank for a time every day in order to imitate the action of the tides. If we keep Convoluta in vessels of water in the laboratory under uniform conditions they will still burrow beneath the surface of the sand at the time when the tide is ebbing from off the beach, although the vessel is kept continually full of water. There are similar rhythms of habits in many other animals. When the sea is phosphorescent in some parts of our waters the luminosity is usually due to a protozoan called Noctiluca, or to copepods, or to some peridinians. All these animals become phosphorescent only when it is dark, that is after sunset. But if we capture them and keep them in a photographer's dark room we find that they become phosphorescent there only when it is dark outside under natural conditions: that is the animals seem to remember the regular alternation of day and night, even when they are kept under strictly uniform conditions. The Chameleon Shrimp, Hippolyte, shows a similar rhythmic habit.

This little animal inhabits sea-weeds of various colours and it nearly always varies in hue so as to match the colour of the alga on which it rests. But in the dark, after the sun has set, all varieties of Hippolyte, whether brown or green or red, become a beautiful transparent blue ; and this nocturnal blue is regularly assumed in the conditions of the laboratory. This periodicity of habit in Convoluta and Hippolyte is fully described in Keeble's volume on 'Plant-Animals' in this series of Manuals.

The annual changes in the temperature and salinity of the sea are of much more importance in their influence on marine life than are the changes of tide. If we consider in the meantime only the higher marine animals, the fishes, molluscs, and crustacea, we can easily make out three phases in their vital condition throughout the year. A phase of partially suspended animation coincides with the time when the sea temperature is about its minimum; then there is the reproductive phase which is assumed in the spring, when the temperature is rising rapidly from its lowest point; and then follows the growth phase which coincides with the rise towards the maximum and the period immediately following. The dark and cold winter months, December to February, are those during which the activities of marine animals are at their lowest ebb. Many of the fishes and crustaceans such as the plaice, flounder and shrimp hibernate, burying themselves beneath the surface of the sand while the temperature of the sea is at its lowest. Nearly all marine animals become sluggish and inert at this time: they do not migrate much; they do not feed so actively as at other times, or perhaps they refuse to feed altogether. Reproduction does not take place, or if it does occur it is mimimal in degree.

What we call functional metabolism, that is the assimilation and oxidation of food for the purposes of keeping up the life-processes of the animal, is reduced to a minimum; the animal uses up as little of its tissues as possible. But morphogenetic metabolism, that is the formation of new substance in the shape of the genital products—the eggs or spermatozoa—increases to a maximum during the resting stage of the cold winter months. During this time the animal is using its own reserve food-stuff, that is, the fats, carbohydrates, and proteids stored in its flesh, for the purpose of manufacturing the generative substances. The ovaries and testes thus increase largely in weight, but the general tissues of the body lose in weight. This of course only occurs in sexually mature animals: if the fish is immature the whole weight of the body falls off. If it is mature the 'condition' is worst of all just after the time of spawning, and the fish is then thinner than at any other time in the year.

The reproductive phase, that is the actual spawning

act in the case of an egg-laying animal, or the birth of
the young in the case of a viviparous animal, almost
always occurs during the months of February to July
in northern seas. The majority of fishes, the cod,
plaice, flounder, whiting, haddock, and many others
spawn during the months of March and April. The
sole spawns during May and June, and the turbot,
brill, and gurnards about the same time, or a little
later in the year. The majority of invertebrate
animals also spawn during March, April and May.
The herring spawns somewhere in British seas
throughout the year, that is we have 'winter' and
'summer' herring. But even here the reproductive
act is a strictly periodic one for the various races of
herring spawning throughout the year are local ones,
each with its own reproductive phase. The act of
spawning lasts for a variable time: it is prolonged in
the case of the flat fishes, such as the plaice or sole,
where the eggs ripen in successive batches, but it
takes place quickly in the case of the round fishes
like the cod, where all the eggs come to maturity at
about the same time. It may be prolonged over a
month or two in the case of the skates and rays where
two large eggs are laid at a time. In the case of the
fishes which lay a great number of eggs—that is in
the majority—the numbers laid at first are few, but
they soon increase to a maximum and then quickly
decrease. Both the time of the maximum spawning

and the duration of the spawning act are variable from year to year. The date of liberation of the eggs depends on the temperature of the water—until the latter has attained a certain point it does not occur. But it also depends on the temperature of the sea during the month or two previous to the time of the actual reproductive act, for the maturation of the generative products proceeds slowly when the temperature of the sea is low, and more quickly when it is high. That is to say the temperature effect is partly an integrative one.

The growth phase begins almost immediately after the completion of the spawning, for in some way or other the parental habit compels the animal almost entirely to neglect its own nutrition. Thus the male lumpsucker, a shore-living fish, attends to the mass of eggs laid by the female, the latter having deserted them immediately on depositing them on the sea-shore. It is necessary for the male parent to 'stand by' the eggs and prevent them from being devoured by other small fishes or invertebrates, and to keep them supplied with a current of water which he sets up by movements of his fins. During all this time he refuses to feed (even if food is supplied to him) while he and the mass of eggs are being kept in an aquarium. But immediately after the eggs have mostly hatched out he begins to feed greedily. This is the case with nearly all fishes

—their time of gorging is after the end of the spawning period. From then till about midway between
the time of the annual temperature maximum and
that of the minimum both functional and morphogenetic metabolism are most intense. The animal
lives at a more rapid pace during the summer months
and takes in more food and more oxygen and gives
off more carbonic acid during that time. But the
type of constructive metabolism undergoes change,
for instead of the development of the reproductive
organs we find that a rapid growth of the tissues takes
place. The fish increases in size, and assimilated
food reserves are stored up in its flesh. All the
growth of the year occurs during this phase, and
the greatest amount of growth takes place at the
time when the temperature is at its highest. Growth
in a marine animal is a strictly 'periodic function'
and is capable of representation by a mathematical
expression. In the case of most marine fishes the
rate of growth can often be represented by a 'damped
sine curve,' that is it is higher than the mean, and
maximal, during the warm months, and it is less
than the mean, and minimal, during the cold months.
The period is an annual one, and the amplitude
decreases as the fish becomes older, following an
exponential law. Such an expression for yearly growth
may also apply to most mollusca, but not to crustacea,
for in the latter animals the growth takes place

by jumps, immediately after the animal has cast its shell. It has happened that a crab has crept into an empty bottle just after casting, and then having grown to its annual amount during the next week or so it finds itself unable to leave the bottle. In the case of animals growing in this way the growth function is a 'discontinuous' one.

Thus some time in the spring months a wave of production of organic substance in the form of new individuals sweeps across the sea, dying away in the summer months. This wave is exhibited in the annual spawning of almost all the higher animals and it affects the composition of the plankton. Just as certainly as the vernal equinox approaches and the sea temperature rises, so does the mass of planktonic animals and plants inhabiting a unit volume of sea-water increase to its annual maximum, and then decrease again. Now we must clearly distinguish between the increase in the mass of the plankton due to the myriads of eggs and larvae spawned by the fishes and larger invertebrates, and that due to the reproduction of the pelagic diatoms, algae, protozoa, and micro-crustacea. The former have only a transitory planktonic phase in their life-histories, the latter persist in the plankton throughout their entire lives.

The transitory plankton includes a great number of species each of which is strictly periodic in its

appearance. The succession of the planktonic larvae
in any sea area is generally the same from year to
year, and it can be predicted. Let us take the succession of organisms in the sea of Liverpool Bay as
an example. We should notice first of all the eggs
of fishes, chiefly cod, plaice, whiting, dab and some
others, and then the larvae of these species of fishes.
By the end of May all these would have disappeared.
The larvae of barnacles appear in the water during
the month of March, with the larvae of numerous
species of crabs, and all these persist for a time, but
soon the barnacle larvae become metamorphosed to
their final stages, and the crab larvae are also transformed to further forms. About the end of May the
barnacle larvae would have disappeared entirely, but
the later crab larvae would still persist. About the
end of May the larvae of shrimps would appear in
quantity, but would very soon leave the plankton, as
they take up their permanent habitat on the sea-
bottom. At the end of May and the beginning of June
the later spawning fishes would contribute their eggs
to the plankton, but by the end of August these also
would mostly have gone down to the sea-bottom as
demersal organisms. June is the season for many other
planktonic larvae, such as those of many marine worms,
sea-urchins, or starfishes, and of these some would last
on till near the end of the autumn, but by that time
most of the larvae would have left the plankton.

We see then that a great number of fishes and invertebrates reproduce in the spring and summer months and that the plankton increases largely in bulk because of the immense numbers of eggs and larvae so produced. But the increase in the mass of the plankton during the spring months—which is so remarkable a phenomenon in the sea—is due far less to the appearance of these larvae than to the reproduction of the pelagic unicellular organisms. Just as each species of fish and invertebrate has its own spawning time, so each species of diatom or protozoan making up the permanent plankton also reproduces at a particular time. The stimulus to the reproduction of the higher animals is the temperature change in the spring of the year, and this has also something to do with the reproduction of the diatoms and protozoa, but it appears that the change of sunlight and salinity are also factors of great importance in the latter cases. During the first two months of the year there are very few diatoms in the sea in Liverpool Bay, but sometimes about the beginning of March certain spring species begin to appear, and attain a maximum of abundance during that and the next month, while at the same time the animal plankton, which was characteristic of the sea before this time becomes temporarily less. The spring diatoms (which are best represented by the two genera Biddulphia and Coscinodiscus) become very

much less in May and June and are succeeded by
others which we may regard as summer forms. In
this part of the sea the summer diatoms belong to
the genus Rhizosolenia and they are always accompanied by enormous numbers of the pelagic alga
Halosphera, a species which forms little mucilaginous
capsules which are easily visible to the naked eye.
During the month of June, and sometimes part of
July, the sea off the coasts of North Wales swarms
with these organisms, and no biological phenomenon
in the Irish Sea is more remarkable than this in-
vasion. The diatoms and the other algae can easily be
seen with the naked eye in a small quantity of water
from the beach—even so little as will lie in an empty
mussel shell; and the nets of the fishermen are often
covered with a slimy mass produced by the bursting
of the capsules of Halosphera. Both species are introduced by the inflowing Gulf Stream drift which is
strongest along the west coast of England and Wales,
and they reproduce when the temperature of the
water is rising rapidly. Some time in the month of
July both species disappear almost entirely and then
the sea is invaded by shoals of jelly-fishes, belonging
chiefly to the species Aurelia and Rhizostoma, but by
the end of the summer these too become much less
in number and during the autumn they sink to the
bottom and die. Shortly after they have attained
their maximum of abundance the sea becomes

invaded by the protozoan Noctiluca, and about this time the water becomes brilliantly phosphorescent because of the presence of the organism. It lasts well on into the winter, when it disappears. During the autumn, diatoms belonging to various species may again become abundant, though this is not an invariable occurrence, and sometimes during the same period peridinians belonging to the genus Ceratium attain a maximum of abundance.

This series of changes in the planktonic life of the sea is of course characteristic of this particular locality alone and the sequence may be modified in other parts of the northern seas. But almost everywhere the general scheme will be: (1) a relative scarcity of all kinds of life in the first two months of the year; (2) the appearance of various species of diatoms which attain a maximum of temporary abundance, and then become much less in number; (3) a short scarcity of planktonic life during the early summer months; and (4) a relative abundance of various species of protozoa and micro-crustacea during the autumn. As the winter approaches all forms of life again become scarce in the sea.

Now associating these various changes with the changes in the physical conditions of the sea we notice that they are always to be regarded as dependent on certain physical rhythms: the inflowing

water of the Gulf Stream drift which attains its
maximum in the early summer ; the increase in the
intensity of the sunlight in the early spring, and the
rise of sea temperature in the early summer.   They
are therefore strictly periodic in their occurrence and
if there are fluctuations in the time of appearance of
the inflowing Gulf Stream water, or in the rise of sea
temperature, or in the change in intensity of sunlight,
then the time of appearance of the various maxima of
planktonic life may be accelerated or delayed.   If
there are long-period rhythms in the order of the
physical phenomena which we have considered then
there ought also to be long-period rhythms in the
abundance of life in the sea. This is of course difficult
to prove but where we have fishery statistics it is
quite evident that there are such periodicities.   Only
two or three times during the nineteenth century
have great shoals of herring appeared off the west
coasts of England and Wales, in the estuaries of the
Dee and Mersey, and these times no doubt coincided
with some unusual combinations of physical conditions.
There is a winter herring in the Skagerak, off the
coast of Sweden, and this has been recorded since
the year 895 and has been found to recur at intervals
of about 111 years.

But the movements and reproductive habits of
a marine organism are not entirely dependent on

the seasonal changes in the sea, for heredity has stamped certain habits on its organisation, and even if the conditions were to remain the same there would still be periodicity in the life-processes. This periodicity would ultimately be obliterated if the conditions of life were always to remain strictly uniform, but for a time it would manifest itself. If we keep such a fish as an eel in an aquarium we shall find that as the period of its life comes when it should be migrating down to the sea in order to spawn in the deep water off shore, various changes will occur in spite of the fact that the animal is being kept in water of uniform condition. When it reaches a certain age it ought to be seeking water of great depth, and if it is unable to do this its metabolism becomes disturbed, since it is adjusting itself unconsciously to inhabit water where the pressure is very great, but it still is compelled to remain in the shallow water of an aquarium tank. We find therefore that bubbles of gas begin to form underneath the skin because the pressure of the water is far less than that to which the fish is adjusting itself by sheer force of inheritance.

We must regard heredity as a kind of flywheel which causes an organism to undergo periodic changes: reproductive changes, migrations, growth, etc. Apart altogether from external stimuli the flywheel would still revolve and the organisms would still display periodicity. But we cannot doubt that if the external conditions were to become strictly uniform then seasonal change would also largely or entirely disappear from the life-habits of animals.

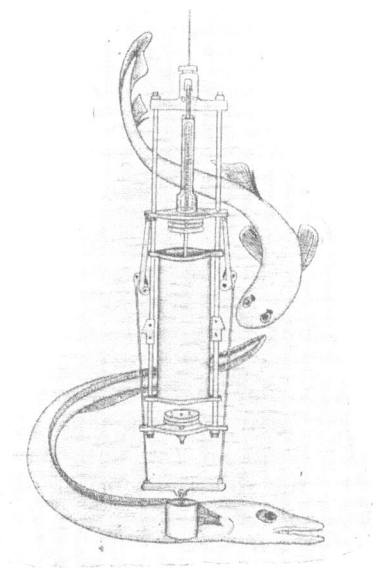

Water-sample bottle and thermometer.

CHAPTER III

THE FACTORS OF DISTRIBUTION

IF we were to compare lists of the species of plants and animals known to occur in the English Channel, the Irish Sea and the Firth of Clyde—adjacent sea areas which have been fairly well investigated, we should find that the majority of the kinds of organisms were common to all three regions, but also that each of the latter possessed a few species which did not occur in one or more of the others. We should also find that certain species were most abundant in one of the areas and decreased in density as we passed north or south. If again we were to compare the fauna and flora of the Indian Ocean off the coasts of Ceylon with those of the North Atlantic between the Faroe Islands and the coasts of Scotland we should find quite the opposite kind of contrast: the great majority of the species would be different in each case, yet there would be a few which were common to the two areas. If, further, we had ex- haustive lists of the animals and plants found in the seas within the Arctic and Antarctic Circles we should find still greater dissimilarity. There would indeed be a general kind of resemblance between the two faunas and floras, but only in the cases of one or two migratory birds and whales would the same species of organism occur in both circumpolar areas. Speaking quite generally, the further apart on the face of the earth two sea areas are, the greater will be the difference between their contained life, because the further apart they are the greater, generally, will be the difference of physical conditions. Very few animals or plants are truly cosmopolitan.

It has been the province of systematic biology to investigate and record the range of geographical distribution of each species of animal and plant, and so far the result has been the accumulation of a mass of detail which possesses the very slightest interest, and out of which it is not easy to extract general laws or regularities. We may, however, try to indicate a few prominent facts. Marine mammalia—the whales, seals, and some other species of warm-blooded animals, are most abundant within the two Polar Circles, and the larger sea-birds inhabit the cold and temperate seas, and are least abundant within the tropics. The marine and anadromous fishes of the sub-polar and temperate seas are far more abundant, and are generally larger than those of the tropics, but the variety of species is not so great as in the latter regions. So also with the invertebrates generally : we find the greatest density in the colder seas but there are more bizarre forms and richer colouring in the warmer waters within the torrid zone. Molluscs are much more abundant in the temperate seas, and generally the shells of the cold-water molluscs are less massive in proportion to the size of the animal than in the warm-water species ; thus the largest marine shellfish—the great Tridacnas and the mother-of-pearl shells—are inhabitants of tropical seas. The pelagic molluscs of the warm waters have generally greatly reduced shells, while those of the northern sub-arctic seas—the winged pteropods on which the whalebone whales feed, have a more massive shell. We find also that those animals which secrete lime do so to a greater extent in the tropical seas : thus the reef-building corals are practically restricted to a belt of sea extending to about 20° north and south from the equator, while even the solitary corals are not very abundant in the colder seas. The micro-crustacea are probably about equally varied in the cold and warm seas but they are vastly more abundant in the temperate and sub-arctic seas. Marine plants are much more plentiful in the temperate and polar seas : the larger rooted Algae are represented by more species and are denser in the temperate seas than anywhere else in the ocean. The pelagic microscopic diatoms, though found everywhere in the seas of the world are conspicuously more abundant the further north or south we go. But some pelagic algae—the small form Trichodesmium, and Sargassum, the gulf-weed, occur in abundance only in tropical waters. The mangrove swamps of the tropical sea-margins are also instances of associations of brackish water plants which do not occur in the temperate or polar seas.

All experience shows that the polar and temperate seas are, generally speaking, far richer in life than are the tropical ones. There is a striking picture of the relative density of life on land and sea in Darwin's description of Tierra del Fuego. That desolate and inhospitable land sustains with difficulty a sparse and cannibalistic population, while the seas near the shore contain such an abundance of life as Darwin could only compare with that of a terrestrial forest in the intertropical zone. Chün, in his account of the voyage of the German exploring ship 'Valdivia,' speaks of the teeming plant and animal life of the ice-covered Antarctic sea. The naturalist Kröyer found an abundance of invertebrate life in the sea off Spitzbergen such as he had never seen surpassed elsewhere. The greatest and most productive fisheries of the world are those of the seas round Britain, off the coasts of northern Norway, on the Banks of Newfoundland, in the seas off Iceland, and off north Europe close up to the ice-packs. Nowhere else in the world has there been such extensive and ruthless destruction of mammalian life as among the whales and seals of Arctic and Antarctic seas. The account of every Antarctic voyage contains striking pictures of the enormous wealth of bird life in the penguin rookeries, and among the petrels and gulls; and also of the incredible abundance of diatoms, which visibly discolour the sea-water and clog the nets used for fishing. All plankton investigation bears witness to the paucity of microscopic life in the tropical seas and to its abundance in the north temperate and sub-polar waters. The German exploring ship 'National' obtained the richest plankton catches in the sea off the south of Greenland, and the poorest in the warm Sargasso Sea; and this has been the experience of most investigators.

We see that there are some general facts of distribution. The variety of species is greatest in tropical seas, and least in the temperate and polar seas; and it is greatest near the land, and least in the oceanic areas far away from the influence of the land: generally the density of both benthos, nektic and planktonic life decreases as we pass out from shallow into deep water. In the middle of the greatest oceans planktonic life is scarcest, and at the bottom of the oceans, in the abyssal regions, the benthos is also scarcest. Perhaps no part of the ocean bed is actually devoid of animal life, but the nearest approach

to the terrestrial deserts is to be found in the great
abyssal plains of the ocean beds. Plant life is most
abundant in the shallow water near the shore, and
least abundant at the surface in the mid-oceanic areas.
It is most abundant just a little way beneath the
surface of the sea everywhere and then it decreases
as we descend into the depths. It is absent entirely
at the bottom of the deep sea.

We have seen also that there are cyclical changes
in the abundance of life of all kinds; changes which
are seasonal ones. Each species is normally restricted
to some particular area of sea or sea-bottom, but
this area expands or contracts with the seasons
and so the range of geographical distribution of the
organisms of the species also expands and contracts;
that is, they are able to migrate. We have now to
consider in what way these general facts of distribution
and distributional changes are to be associated with
the physical conditions in the sea, and how fluctuations of life are to be associated with fluctuations in
the physical conditions.

We consider first of all the physical conditions of
temperature, salinity, and sunlight; and then the
nature and abundance of food in the sea. These are
the factors of distribution. The surface temperature
of the sea is highest a little way north from the equator,
and then it decreases towards the poles, falling at a
nearly equal rate as we pass north or south. This

regular decrease of temperature is modified by the
influence of the land and by ocean currents: thus the
temperature of the Red Sea, and of the Atlantic Ocean
to the west of Ireland are higher than they ought to
be if the latitude were the only factor, because of
the influence of the land in the first case, and that of
the Gulf Stream in the second. The temperature
of the Atlantic off the coast of North America is lower
than it ought to be because of the influence of the
Labrador Current. The temperature of the sea is
very generally highest at the surface and then falls
towards the bottom, quickly at first and then more
slowly. Towards the bottom in very deep seas it may
rise a very little, probably because of the generation
of heat from radio-active substances in the bottom
deposits, but it is a general rule that almost everywhere the temperature at the ocean bottom lies
between 2° C. and zero, and in places it may fall
below the freezing point of fresh water. Occasionally
the deeper or the intermediate layers may be warmer
than those of the surface, but these conditions are
anomalous and are to be traced to the action of
currents.

The salinity of the sea is rather low along a belt
of ocean between the equator and 5° N., that is in the
region of tropical calms and torrential rainfalls. It
is highest along two belts of ocean situated between
25° N. and 30° N.; and between 20° S. and 25° S.,

that is, in the regions of high evaporation due to the
trade winds. On each side of the trade-wind belts
the salinity falls—rapidly towards the north, and
less rapidly towards the south. In the Arctic and
Antarctic seas the salinity is low because of the
melting of ice which contains less salt than sea-water.
Here and there we find marked deviations from this
general distribution of salinity—the Red Sea, for
instance is salter than it ought to be because of the
great evaporation due to the high temperature; and
the sea off the coasts of Denmark at some parts of
the year is fresher than it ought to be because of the
influence of the outflowing Baltic fresh current.

The intensity of sunlight is greatest on the surface
of the tropical seas, and least on that of the circumpolar seas. But the amount of solar radiation falling
on the surface of the sea is greatly affected by the
amount of fog and cloud, and variations due to these
causes affect the abundance of plant life at the surface
of the sea.

The physical conditions over the oceans are far
more uniform than over the land. Thus the extremes
of temperature on the land are $-90°$ C. and $65°$ C., a
range of $155°$ C., while the extremes in the sea are
only $-2.8°$ C. and $31°$ C., a range of $33.8°$ C. There
are no changes in the sea corresponding to changes
of humidity, or to rain, hail, snow, etc., on the land.
As we ascend the high mountains the temperature

falls to a much greater extent than it does as we descend into the depths of the ocean abysses. If we ascend to the summit of a mountain 24,000 feet high the pressure does not decrease by nearly so much as one atmosphere, but as we descend to a corresponding depth in the sea the pressure rises to about 700 atmospheres. The extremes of salinity variation in the sea are about 30 and 36 except in very exceptional cases. The annual change of temperature is much less on the land than in the sea. The annual change of sunlight intensity due to the seasons is of course the same in each case, and so also is its variation from place to place since this depends on the latitude.

The distribution of plants and animals on the land is restricted by mechanical barriers, by mountain chains, deserts, seas, great rivers, etc. In many cases we find that these barriers are insurmountable so that the great continents and the oceanic islands are characterised by the possession of species peculiar to themselves. Birds and insects may fly, or be blown, for great distances, and plant-seeds may be transported in several ways; but we find that the barriers have operated in producing the present distribution and diversity of life; and that apart from human agencies very few species of plants or animals would be truly cosmopolitan. In the sea, however, there are no such mechanical barriers to distribution, for the watery medium is everywhere continuous: on the other hand the oceanic currents afford a powerful means of transport of both animals and plants such as no agency provides on the land.

So we find that relatively large areas of sea or sea-bottom may possess a fairly uniform fauna or flora—the Sargasso Sea in the tropical Atlantic, the intermediate layers of the North Atlantic, or the bottom of the same ocean at over two thousand fathoms depth are examples. Even apart from such cases of wide distribution of a nearly similar life there are cases of the range of isolated species over very large tracts of ocean. Thus some copepods found in the sea off the coast of Ceylon or in the Gulf of Guinea are also found in the waters of the Faroe-Shetland Channel; and some diatoms inhabiting Chinese seas are now known to occur over many parts of the North Atlantic. Peridinians occurring in the Pacific off the coasts of California are also found off the coasts of Norway. Some species of tapeworms inhabiting the intestines of fishes in the sea off Ceylon also infest British fishes. In these cases the agency of the wide limits of distribution is the oceanic circulation—the Chinese and Ceylonese diatoms and copepods have been taken into the Equatorial Stream circulation and so into the Gulf Stream system; the Pacific peridinians have discovered the north-east passage, and have drifted across the North Polar basin; while the tapeworms probably in their early larval stages inhabit some pelagic animal which is drifted for great distances.

The barriers to the universal distribution of marine organisms are to be found in the different types of structure evolved by them, and in corresponding specialisation in the modes of metabolism acquired during the course of their evolution. Most organisms have developed types of bodily conformation which necessarily restrict them to definite areas of sea or sea-bottom. A shore alga lives attached to a stone on the beach and cannot exist floating permanently at the surface of the sea; on the other hand a pelagic alga or diatom lives drifting at the sea-surface and cannot always inhabit the sand or mud of the shore. A pelagic fish may roam over extensive tracts of sea-surface, but if it attempts to descend towards the bottom the increasing pressure will be fatal to it; and conversely an abyssal fish might migrate over the entire ocean bottom of the world sea, but it cannot approach the surface without great distension of its swim-bladder, protrusion of its stomach through its mouth, and disintegration of the muscles and other tissues—such injuries as would be fatal to it. Sessile benthic animals like zoophytes, barnacles, or sea-anemones must live on the beach or at the sea-bottom attached to stones, and cannot drift about at the surface; while littoral animals such as the cockle or lugworm must restrict themselves to the shore

between the tide marks. On the other hand planktonic organisms like medusae or ctenophores must
drift about at the surface, or just beneath the
latter in clear water. If they are stranded by tides
or gales—a thing that often happens—they perish.
In the course of their evolution of certain types of
structure and habit most marine organisms have
surrendered the capacity of existing except under
very restricted conditions.

Yet this limitation of habitat is partially com-
pensated for by the evolution of larval stages in the
life-history of an organism. We nearly always find
that a sessile benthic animal has evolved a free-
swimming larval stage : or the primitive pelagic form
has evolved a sessile habit during the latter period
of its life-history. Whichever view we take it is the
case that the great majority of animals which live
attached to the sea-bottom, or burrowing in the sand
or mud, or crawling or swimming along the sea-
bottom, begin life as eggs which drift about in the sea,
or which may be attached to the parent during their
period of incubation. In either case the creature
which hatches out from the egg-shell is a larva, and
it lives pelagically among the plankton drifting over
relatively wide sea areas. This is the meaning of
a larval stage in the life-history of an animal—
it is an interruption of the course of the development to afford the incompletely formed embryo the

opportunity of feeding and being drifted about by
the tides, winds and currents over as wide an area as
possible. After a pelagic phase of variable duration
the larva undergoes one or more metamorphoses and
assumes the form of the parent, and then settles
down on the sea-bottom. The larva is usually locomotory though the adult is not: thus the rooted
immoveable algae produce motile zoospores and the
fixed zoophytes free-swimming medusoid larvae.
But whether or not the larva is locomotory it is so
small, and its specific gravity is so near that of sea-
water, that it is carried for great distances by tidal
streams and currents, and it may settle down very far
away from the place where it was born. Its life as
a larva is usually a very limited one—if the free-
swimming stage lasted long enough there would not
appear to be any limit to the normal distribution of
the species to which it belongs. The object of the
development of the larval stage—if we may speak of
an 'object' in an evolutionary process—is obviously
the enlargement of the area of distribution of the
species.

This distribution of a sessile, or semi-sedentary,
species by means of free-swimming larvae must be,
to some extent, a fortuitous process. For though
the larvae may appear to be very active creatures
when examined in a few drops of water under a
microscope, yet their powers of locomotion are very

feeble when compared with the distances to which
they may be carried by wind-drifts and currents.
They can have no control over the direction in which
they are being carried, and when the metamorphosis
occurs—a process over which also the larva has no
control—it must assume the habitat of the parent ;
and if it is a benthic form it must settle down on the
sea-bottom.   If it is a littoral form—a cockle or
barnacle for instance—and if the eggs or larvae have
been transported into deep water, when the metamorphosis occurs the newly developed adult will
settle down to the sea-bottom and will be destroyed.
Or the larvae of a pelagic fish inhabiting deep water
may be carried into the brackish water at the mouth
of an estuary and will so be lost.   It is easy to see
that very great numbers of planktonic larvae must
therefore fail to find their proper habitat at the time
when the transformation occurs, and this is one of
the reasons why most marine organisms produce
such large numbers of pelagic eggs—the turbot,
for instance, may spawn as many as nine millions of
eggs each year ; even the relatively slowly reproducing barnacle spawns about five to nine thousand
eggs.

    Yet the manner of distribution is not such a
hap-hazard process as it might appear to be, for
though the majority of larvae may fail to enter
adult life at the proper time or place, enough may be

successfully 'planted' to carry on the species in un-
diminished numbers. Moreover the eggs are usually
so liberated by the parent that the currents or tides
will carry them to suitable places. If the parent is
sessile in habit it is usually found in shallow water
near the shore, and the tidal streams flow, as a rule,
along the land, or surge backwards and forwards to
and from the land ; therefore the larvae may inhabit
shallow water during the whole of their free-swimming period. If the parent animal is a nektic one it
usually performs a spawning migration, that is to say
it migrates when about to spawn to such a place as
will ensure that its eggs will not all be lost. The
majority of fishes produce buoyant eggs which float
at or near to the surface of the sea, but the larvae
hatched from these eggs prefer to inhabit shallow
water near the shore. When the fish is about to
spawn it migrates out to sea to some distance from
the land and then spawns, and the spawning place is
so selected that the currents at that time of year, or
the prevailing winds will drift the eggs towards the
land. The cod, plaice, flounder, sole and many other
fishes behave in this way—they migrate offshore to
spawn, and then the eggs drift inshore, completing
their embryonic development during the transporta-
tion, so that, barring accidents, the larvae are just
about to undergo metamorphosis by the time they
reach shallow water. The spawning takes place

during the spring when there is abundant food, and when the temperature is rising rapidly in the shallow water near the land ; therefore the young larvae are carried to a land of plenty. The egg-laying skates and rays behave in an analogous manner, migrating towards parts of the sea where the bottom is rough and stony and contains abundance of sea-weed, and where there is little run of tide. The eggs are laid on the sea-bottom where they develop without much risk of being washed away by the tide, or stranded on muddy beaches, or taken out into deeper water. The herring makes a somewhat similar migration for the same purpose. The salmon was probably originally a marine fish which adopted the habit of migrating into rivers in order that its eggs might not be exposed to the same risks of destruction by such enemies as they would encounter in the sea. Thus the spawning migrations of fishes are directed movements carried out to achieve a definite purpose. We need not suppose that they are conscious actions ; and we may regard them as 'instinctive acts.' The stimulus to the performance of the spawning migration is the development and ripening of the generative organs, and the elaboration of an internal secretion from the ovary or testis which produces an intoxication, and impels the fish to seek water of definite physical conditions. What these conditions may be depends on the former history of the species—the 'historical basis of acting' being the determining factor in this choice.

Thus the evolutionary history of a species has determined to some degree its mode and extent of distribution; for the development of definite structures and habits restricts the animal to a particular range of habitats. Nevertheless it is easy to see that most species of marine animals would be distributed over a much wider area than they actually are if there were not some factors which limit this distribution. These factors are those of temperature, salinity, pressure of water, and intensity of sunlight. We have next to consider in what way these affect marine organisms.

The temperature is probably the most important of all these factors, and it is easy to see that there are many ways in which a marine animal, such as a fish, can be affected by changes in it. There is no reason to suppose that a fish cannot feel changes of temperature in the way we do, that is by its sense-organs. It appears that sudden changes can produce pathological effects, for fresh-water fish are known to be affected by a form of dermal catarrh when they are suddenly moved from water of high to water of much lower temperature; and there is a record of the destruction of millions of tile-fish off the coast of North America by a rapid and extensive chilling of the water stratum in which they were living.

Such sudden temperature changes must however be very exceptional in the sea, and a nektic animal, such as a fish, would usually be able to respond to them by means of a migration into water of more favourable conditions. Temperature changes must have for their usual effects changes in the intensity and mode of metabolism of the animal; and there is plenty of experimental evidence that this is the case. We have to regard the life-processes of an animal as the result in some way of a regulated series of chemical reactions, and we know that the velocity of a chemical reaction increases as the temperature rises. The amount of oxygen absorbed in respiration and that of the carbonic acid excreted are greater or less according as the temperature is greater or less; and the rapidity of respiration varies in the same way. In cold water the respiratory movements of the mouth and gills of a fish slow down greatly and often assume the Cheyne-Stokes type, that is there are relatively long pauses followed by groups of respirations. This means that the rate of metabolism, that is the amount of chemical change going on in the tissues of the animal, decreases as the temperature falls. Towards the end of the year the quantity of food taken decreases and many fishes such as the plaice and flounder cease altogether to feed during the months of December to February. The stomach and intestine are usually empty during this period;

tissue waste is reduced to the minimum ; and the
animal hibernates by burying itself in the sand at the
sea-bottom.   The tissues are however wasted away
to a greater extent than they are renewed, and the
'condition' of the fish is at its worst at the period of
minimal sea temperature.

The rate of development of the embryo of a
marine animal varies also with the temperature ;
thus a flounder embryo will hatch out from the egg
in about ten days at the beginning of the spawning
period, when the water is coldest ; but it will hatch
out in about six days at the end of the spawning
period when the water has become much warmer.
This relation holds for all developing eggs of cold-
blooded animals which have been investigated, and
it has been stated as follows ; for a difference of
temperature of 10° C.

$$\frac{\text{Average duration of development at } n° \text{C.}}{\text{Average duration of development at } (n-10)°} = \frac{2\cdot85}{1}$$

which means that when the temperature is reduced
by 10° C. the time required for the hatching out of
the embryo of a marine animal is increased about
threefold.

The duration of life also varies with the temperature in such a way that the lower the temperature of
its life-medium the longer an animal lives.   Probably
this also holds good for warm-blooded animals—for

instance cats which live in cold-storage warehouses
have sleeker coats and are apparently healthier than
are animals living under normal conditions ; and it
is possible that the connection between length of life
and temperature might also be proved by comparative
vital statistics.   However this may be, the following
experiment affords direct evidence of the probability
of the statement : sea-urchin eggs are fertilised and
are kept at a constant temperature, and then small
quantities are taken and are put into water of
different temperatures, and the higher the latter the
shorter is the life of the egg.   If we call the duration
of life of an animal at temperature $t°$ a_t, then the
duration of life at temperature $(t - n)°$ is $a_t 2^n$.   This
means, supposing of course that the experiments
referred to should be confirmed, and prove to be
of general application, that if the temperature of the
water in which a cold-blooded animal is living is
reduced by 10° C. its duration of life will be increased
a thousand-fold.   It seems too good to be true !

It would seem, however, that some such relation
between temperature and length of life of the kind
stated does actually exist.   Now let us apply it to
the case of the observed density of plant and animal
life in warm and cold seas and we have a possible
explanation—not the only one as we shall see—but
still perhaps the most probable one.   When the
temperature is reduced by 10° C.—not so great a

difference as we find between tropical and polar seas, the duration of life is increased about a thousand-fold, but the rate of development is reduced to only a third. This means that a marine organism will reproduce about three times less actively in cold seas than it will in warm seas, but it will live a thousand times longer ; and therefore it must be much more prolific and there will be many more overlapping generations in the polar than in the tropical seas. If the relation holds good we *must* have a greater density of life in the former waters than in the latter. We should not, indeed, expect to find such a difference as would be expected from the experiment described, for the abundance of food would limit the development, and so also would other factors—the less intensity of sunlight, for instance, in the case of a marine plant, such as a diatom. But it seems clear that the difference between rate of development and duration of life is a factor. How exactly the temperature affects the duration of life we do not know : it may be that some substances are produced, and are accumulated in the tissues, and that these substances are detrimental to life. If these toxic substances are produced at a greater rate the higher is the temperature, then the duration of life will be shorter at high than at low temperatures.

How does a change in the salinity of the sea-water affect a marine animal ? If a fish, say, moves from water containing a relatively high concentration of salt to water containing a low concentration, can it perceive the change by means of its receptor taste organs ? It does not seem likely that we could perceive by taste alone the difference between water containing $3·45\,{}^0/_0$ of salt and water containing $3·40\,{}^0/_0$, nevertheless this is a change towards which a fish might react. Now the atmosphere is to us very much what the sea-water is to a fish, and we can detect by means of our sense-organs changes less than that instanced. Let us suppose that the temperature falls rather suddenly from $20°$ C. to $15·6°$ C.: we should certainly be able to detect such a change. It would be a change of $1·5\,{}^0/_0$ of the total possible range of temperature, regarding $20°$ C. as the upper limit; and the salinity change is also $1·5\,{}^0/_0$ of the total range, regarding $3·45\,{}^0/_0$ as the upper limit. It seems quite probable then that the fish would be able to detect, by means of its taste-buds, the change in the saltness of the water, especially as we have reason to believe that the olfactory receptor organs in a fish have a lower threshold than in our own case, and smell is only 'taste at a distance.'

But we are not restricted to the supposition that the fish can only be affected by salinity changes if it can appreciate those by its organs of taste. Its blood and lymph are watery fluids containing certain salts of sodium, potassium, lime and magnesium in

solution, and these liquids are separated from the
surrounding sea-water only by a membrane which is
very thin in certain places, over the gills for instance.
Now the sea also contains these same salts in greater
concentration than they are contained in the blood ;
and if the membrane covering the gills were a dead
one, then water would diffuse out from the blood and
lymph of the fish into the surrounding sea-water.
It does so diffuse but only to a very slight extent, and
not nearly so much as would be expected if the
process of diffusion in the case of a living membrane
were the same as that which holds good for an
inorganic one. Diffusion does not take place to the
full extent possible, for among all the chemical and
physical reactions that may occur in the tissues of
an animal some do not happen because of the power
of regulation possessed by the living organism. But
if the concentration of the sea-water is higher than
that of the blood, and if diffusion does not take place,
then work must be done to prevent it. If the fish
moves from a region of high to one of low concentration less work must be done, and this is a reaction
on the part of the organism towards the change of
salinity.

It is easy to show that changes in the intensity
of sunlight must directly affect a marine organism.
They do, of course, profoundly affect a green plant,
as we shall see later ; but animals which do not

contain chlorophyll in the tissues are also affected.
Those which have eyes will probably perceive light
changes in the way we do, but even if eyes are
absent there is always the possibility that light
changes can be perceived by means of the skin.
Intense light radiation may even kill some of the
lower organisms, and in ourselves certain well-known
curative effects depend on some such reaction on
the part of diseased tissues.　Almost all animals
which can be kept in aquaria respond by directed
movements to changes in the intensity of the light
falling on them.　In the cases of the fishes and
higher crustacea these movements are usually away
from the source of light, but in the cases of the lower
organisms they may be either towards or away from
the light.　Plaice and flounders in aquaria usually
prefer the darker parts ; and in the sea they lie on
the bottom during the day, but at night they may
swim up towards the surface.　Crustacean larvae
usually swim to the source of light when the other
conditions are normal.　The abundance of animal
plankton in the sea is affected by the intensity of
light in that the organisms move up and down from
stratum to stratum of water.　We usually get a
different kind of plankton during the day from that
we get at night, and even the amount of cloud in
the sky will make a difference.

　　If a species of plant or animal can possibly

inhabit a wide area of sea, but is restricted to only
a part of that area it is because it finds in the area
which it has come to inhabit its *optimal conditions*
of life. In the course of its evolution not only has
a definite form been developed, but also an adaptation to certain conditions of temperature, depth of
water, salinity, and sunlight. It does not follow that
because the optimal temperature for a marine diatom,
say, is 10° C. it cannot exist at 0° C. or 15° C., but
its vitality is greatest at 10° C., and it reproduces
most accurately at that temperature. If we take
some beef tea and inoculate this with some species
of bacillus we shall probably find that the most rapid
reproduction takes place at a certain temperature—
37° C. may be a common optimum. Far below this
temperature the bacillus would still be able to live,
but it will not reproduce when the temperature
approaches the zero of the Centigrade scale, and at a
temperature not very far above the optimum it will die.
Most species of unicellular organisms are able to live
at any temperature found in the sea and some are
said to be able to withstand even the cold of liquid
air ; almost any salinity found in the sea is fit for
marine life, and any degree of ordinary illumination.
But the rate of reproduction of any organism is
greatest for a certain combination of all the physical
factors, and if these are varied, or even if one of
them is varied greatly, the rate of reproduction is

affected favourably or unfavourably according to
the direction of the variation. If the latter is great
enough the species will not reproduce at all even
if it should still be able to live. At all times the
organisms of a species are subject to destruction by
their natural enemies, and the species is only able
to maintain itself if its powers of reproduction are
at least equal to the rate at which it is being
destroyed. It will not usually be the case that a
species lives under optimal conditions : generally it
will just be able to hold its own against its enemies
but at certain times and places the conditions become
optimal and then the species attains a temporary
maximum of abundance. Such a maximum occurs
during the spring in the case of the diatoms, for the
temperature, the food, and the intensity of sunlight
have then become optimal.

Where in the sea the conditions are optimal there
the species has its centre of distribution. From this
centre it spreads out in all directions that are possible,
for the tendency is for it to enlarge its area of distribution, and the further we go from the centre the less
abundant it will be, and at a certain distance it will
disappear entirely. Near an imaginary contour line
drawn round the centre of distribution the rate of
reproduction is just equal to the rate of destruction;
beyond this line the latter becomes the more powerful
of the two, and the species becomes rarer and rarer

until it can no longer be found. The area within the contour line we may call the area of productive distribution, that without it, and including all parts of the sea where the species has occurred, we may call the area of nominal distribution. It is unable to maintain itself within the latter area unless it is recruited from the former one.

The individuals of a species may be carried far beyond their area of optimal distribution, for if pelagic larvae are produced they may be transported for considerable distances by currents. Thus the pelagic larvae of the crustacean Squilla may be found in St George's Channel but we have no record of the occurrence of the adult there. It probably reproduces only in the Biscay-Mediterranean waters and the larvae are carried to the north in the Gulf Stream drift. So also with the copepods which are occasionally found in the Faroe-Shetland Channel: their productive area is in tropical or sub-tropical seas and the larvae are simply drifted into northern waters, growing but not reproducing there. If such pelagic larvae were to live long enough there would probably be no limit to their area of nominal distribution. Powerfully swimming nektic animals or migratory birds are also able to migrate far beyond their area of optimal conditions; thus sturgeon, breeding only in the great rivers flowing into the Baltic may nevertheless be found over practically the whole north

European seas; and the Arctic tern, which breeds only in the Arctic Circle may yet be found in Antarctic seas.

But the area of optimal conditions varies with the seasons and we therefore find that the productive distribution of most species fluctuates within certain limits. Flounders live under optimal conditions in the brackish waters of the estuaries when the temperature of the water there is higher than it is in the open sea offshore; but whenever the first winter snows begin to melt and form freshets in the rivers then the fish begin to migrate offshore, where the temperature is now higher than it is in the estuarine areas. The cod which come into British waters during the winter have migrated down from more northerly latitudes because when the temperature of the sea was falling there it was a little warmer to the south. The sea-perch, mackerel and gar-fish (Belone) are species which have their area of productive distribution to the south-east of the British Isles, in the Mediterranean, and off the coasts of southern Europe; they migrate into northern waters because the flooding of the latter with Atlantic water in the spring and the seasonal rise of temperature have enlarged their area of distribution. Plaice from more southern latitudes invade the Barentz Sea to the north of Europe in the autumn because the temperature of the bottom water there has been raised by the inpouring of water from the

Gulf Stream drift. If we study the migration paths of some species of fishes we can see they move so as generally to cross the isohaline or isothermal lines in the sea. If in any place the salinity and temperature alter, the fish seem often to move so as to regain the former conditions. Many fishes inhabit, by preference, not the inflowing Atlantic water of high salinity, nor the fresher coastal water, but rather the mixture of the two. The herring which are caught during the summer off the east coasts of Britain first appear off the Western Hebrides and are then found further and further to the south as the season advances. This used to be explained by supposing that the fish migrated from somewhere in the north and then moved to the south in the course of the season. But the herrings which appear at each station spawn, and it can hardly be the case that they all belong to the same shoals. It is far more probable that there are local races of fish which inhabit the North Sea throughout the year and that they shoal and spawn when the conditions become optimal. This optimum of conditions is the result of the inflowing Atlantic water from the Gulf Stream drift, and this flows further to the south as the year advances, while the herring cling to the borders of the salt water where it is mixing with the fresher water coming from the land. Halibut and ling inhabit the bed of the North Sea where it slopes down into the depths of the Norwegian Sea. At the bottom of the latter is a stratum of cold Arctic water and overlying this is the warmer and salter water from the Gulf Stream, while between these strata is a layer intermediate between the two in character, and this is the region of optimal conditions for the fish. Each of the principal strata of water shifts periodically with the rhythmic movements of the Gulf Stream and Arctic current, and the fish migrate so as always to remain in the mixture of water from the two sources.

A bottom-living, sessile animal is unable to respond to changes in the physical conditions by migrations; it can only do so by changes in its own mode of metabolism. It can, however, distribute itself over a wide area by reason of the fact that it produces free-swimming larvae which are carried in the currents of the sea. But a nektic organism which usually also produces free-swimming larvae must generally have a wider range of distribution than a bottom-living fixed one, since it can migrate so that it will always remain within the area of optimal conditions. It is usual nowadays to describe the movement of an animal as a 'tropism' or 'taxis.' The tropistic theory is the result of the modern tendency to look upon all the activities of an organism as the results of known chemical or physical laws. A taxis is some movement on the part of an organism which follows inevitably on some external stimulus, unless it is prevented by some other stimulus, or by some regulatory action effected by the nervous system of the organism. The flight of a moth into a candle flame is an example of phototaxis, and so is the action of a bird when it dashes itself against the glass of a lighthouse lantern. Now probably these are acts which can fairly well be described as inevitable taxic movements which the animal must perform under the irresistible stimulus of the light. So also the movement of an inebriate towards a public-house door, when he had just been given a sixpence, would fittingly be described as an inevitable taxis or tropism. But it seems to be rather straining after generality to attempt to describe all the movements and migrations of organisms as tropisms. This, for instance, is the kind of physiology that is often taught. A certain caterpillar feeds on the tender shoots at the upper extremities of a plant, and to reach these it must migrate upwards towards the light, for the plant grows so that it always bends towards the source of light. The caterpillar is said to be 'orientated' by the light falling on its bodily surface and it must place itself so that the light strikes equally on both sides of its body. Since it moves upwards it is said to be 'positively phototactic.' But having fed it now moves downwards again and it is now said to be 'negatively phototactic,' the act of feeding having reversed its reaction to light. Obviously one might describe the movements of a hungry man

6—2

towards and then away from a fried-fish shop in just the same way. Now there is another manner of description which will appeal with greater force to some minds, and which makes use of effective consciousness as an agency of regulation of the movements of an organism. An animal in good health has superabundant vitality, and it moves about continually in all directions, it may be in search of food, it may be quite capriciously. It is adapted to live best in certain optimal conditions and if it happens to move into a region where the conditions are unfavourable it will endeavour to avoid this region. Its method is that of 'trial and error.' For instance, plaice of over one and under two years of age have for their optimal conditions those of the sea in shallow water where the temperature is rather high, where there is plenty of light, and where there is plenty of food in the shape of small shell-fish. As the fish moves about it must often enter the region of darker, colder, and deeper water offshore, but having done so both the decrease in the intensity of light, and either the increasing pressure of the water or the increase of salinity will affect it in some of the ways suggested, and then it will turn this way and that, testing the conditions all round it by means of its receptor organs, until finally it reverses its former path and re-enters the former region of optimal conditions. If a rapidly growing bed of young mussels forms anywhere within the area

of optimal conditions of such young plaice, it will very
soon be inhabited by a multitude of fish feeding
greedily on the nutritious shell-fish. Now we must
assume that they actually 'like' the mussels, for it is
the case that even if other food is abundant the plaice
of certain regions will take the young mussels to the
exclusion of other shell-fish—there must therefore be
choice on the part of the fish. A fisherman will
doubtless say that the bed of mussels has 'attracted'
the plaice, and a physiologist of the mechanistic school
will explain the aggregation of the fish on the shell-fish
bed in terms of his ingenuous theory of tropisms, and
may call it an example of 'trophotaxis,' or some such
term. What doubtless happens is that the fish,
moving about in all directions within their optimal
area, accidentally encounter the shell-fish bed and
finding food which is to their liking they will stay
there. While feeding upon the mussels they still
move about and it must often happen that a fish will
leave the bed and so will not obtain so much food.
But by turning about it will again, after a number of
trials and failures, re-enter the area of abundant food.
In attempting to explain the movements of a marine
animal we assume the existence of consciousness in
it, for the same reasons as we have to assume it in
other individuals of our species; and it seems sound
enough to assume that in the fish as in ourselves consciousness has become an effective factor of evolution

and behaviour. We must assume the quality of
memory on the part of the animal—it may be some
kind of unconscious memory. Then it follows that
an animal will react not entirely in response to the
stimulus which reaches it from without, but also in
response to all the stimuli which have affected it in
the past; that is to say its behaviour at any time is
modified by its experience. Now if one attentively
studies the facts of migration it seems that this way
of looking at them is more in accord with what one
sees than simply to suppose that the behaviour is
a series of inevitable responses to changes in the
environment.

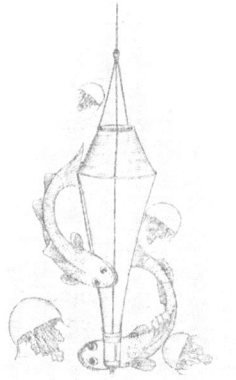

The quantitative plankton net.

CHAPTER IV

MODES OF NUTRITION

It is thus easy to show that the life-processes of marine organisms are directly affected by changes in the external conditions under which they live: that is to say in the chemical composition of the sea-water; in the temperature of the sea; in the intensity of solar radiation other than heat; in the pressure of the sea-water and so on. But each species of organism feeds in some manner and its density and range of geographical distribution must be affected by changes in the abundance of its food. Now most marine animals nourish themselves by eating other animals or plants; and most marine plants by assimilating certain substances which are contained in solution in the water of the sea. We have therefore to consider what are the precise ways in which these nutritive processes are carried out, and how the food-stuffs are distributed in the sea.

The majority of marine animals are predatory, and the history of the evolution of their bodily form is to

a great extent that of the development of organs for
the capture of food. Such organs are the jaws, teeth
and some other structures in fishes; the suctorial
mouths and rasping tongues of lampreys; the suckers
and horny beaks of the squids and cuttle-fishes, the
pincers, maxillae and mandibles of the crustacea;
the blood-sucking mouths of leeches and some other
worms; the scraping tongues, or radulae, of whelks;
the tentacles and stinging cells of sea-anemones and
jelly-fish; the suckers and protrusible stomachs of
starfish and sea-urchins, etc. These are examples
which illustrate the great diversity of methods by
which the predatory marine animals capture their
prey. The majority of nektic and benthic animals
hunt for the organisms which serve them as food,
even the microscopic amoebae and infusorians appear
to roam about incessantly in search of food. Many
marine animals are cannibalistic, and not a few of
them are quite promiscuous in their choice of food,
but as a rule each species eats by preference one or
a few species of food organisms. A small minority
of the nektic animals, chiefly the whalebone whales,
and some pelagic fishes like the herring and mackerel,
are plankton feeders. The whales take in mouthfuls
of water, and then on closing the jaws this water is
forced out through a sieve formed by the frayed-out
margins of the whalebone plates. The herring and
mackerel take water into the mouth and then the

latter is closed, but because of the presence of mandibular valves this water is forced out through the gill-apertures and is compelled to pass through a sieve formed by delicate bones on the hinder margins of the gills—the gill-rakers—and the particles of food are thus intercepted. In each case, that of the whale and the pelagic fish, the food is scraped off by means of the tongue and is then swallowed. Sedentary shell-fish like the cockle or mussel cause a current of water to flow through the shell cavity by means of ciliary action. This water passes through a fine filter formed by the gills and the sediment and plankton contained in it are thus intercepted and are carried into the mouth by means of another system of cilia. Sedentary tunicates have a similar manner of feeding. Sponges act in the same way, by causing a current of water to flow in through the pores and then out again through the osculum. In traversing the system of canals which permeates the body of the sponge the food substance is removed from the water and is taken up directly by the cells lining these cavities or canals. Barnacles open their shells at regular intervals and protrude a bunch of appendages known as the 'glass hand': these are finely fringed by hairs so that when they open out they form an apparatus like the 'cast-net' of a fisherman, and this, as it sweeps through the water, captures any small organisms contained therein. When the hand

is withdrawn into the shell the food particles are
scraped off by other appendages and are then
swallowed. Some microscopic infusorians, such as
the Vorticellids, which are attached to fixed objects,
cause a current of water to flow towards the oral
funnel or mouth of the creature, and this current
carries with it planktonic food organisms or food in
some other form. Other marine animals have more
complex methods of capturing their food, but these
examples will illustrate the variety of means employed.

The majority of the larger marine animals take
the food into their alimentary canals and digest
it in essentially the same way as does the warm-
blooded mammal. Many fishes crush the shells of the
molluscs or crustacea which they use as food between
the hard roof of the mouth and the bones which form
the skeleton of the gills ; crustacea, as a rule, hold
the food animal in their pincers and then pull it to
pieces by means of the masticatory appendages ;
squids and cuttle-fishes crush the food in their
tentacles and then disintegrate it by the horny beaks ;
and many molluscs scrape away the food by their
rough tongues or radulae. The teeth and jaws of the
fishes are organs for the prehension of the food
rather than for its trituration, and these animals, as
a rule, 'bolt' their food. Starfishes ingest and
digest their food outside their bodies, for they can

hold a mussel, say, and then by means of long-continued traction they cause the latter to open its shell when the starfish protrudes its stomach and sucks away the soft parts of the mollusc. In the larger animals there is however very little trituration of the food in the mouth and hardly at all any insalivation.

The processes of digestion and assimilation are probably the same in the fishes and invertebrates as in the mammals, but one must not be too dogmatic on this point for very little exhaustive investigation has been made with regard to the digestive processes of the larger invertebrata. The broken-down food contains, in addition to its indigestible skeleton the 'proximate food-stuffs' of the dietetic text-books ; that is proteid (which is the substance of flesh), fats, and carbohydrates (starches, sugars and gums). The general composition of the food of the marine animals is therefore similar to that of the terrestrial carnivores but it must usually contain a larger proportion of non-assimilable substance, in the shape of the calcareous or horny shells and carapaces of molluscs and crustacea. The herbivores of the land are represented in the sea by a few fishes such as the mullets which 'graze' on the alga-covered stones of the sea-bottom, and by the plankton feeders. The food of the pasture-feeding or frugivorous land animals contains a relatively large proportion of

difficultly assimilable cellulose, and that of the
herbivorous marine animals contains in place of
cellulose the calcareous or siliceous shells or skeletons
of the food animals. Generally speaking the food
of the marine carnivores and herbivores corresponds
to that of their counterparts on the land, but bulk
for bulk their food is not so rich in the tissue-forming
constituents, nor in the heat-producing fats or carbohydrates.

The processes of digestion must be clearly
distinguished from those of assimilation—an important distinction which is made far more clear by
the vegetable physiologists than by writers on marine
biology. The tissues which compose the bodies of
plants and animals are made up of cells modified in
innumerable ways to carry on vital processes.
Organic cells form about them many other structures :
fibrous structures, calcareous and siliceous structures
in the form of shells, etc., and other hard parts ;
they also lay down in themselves substances such as
fats, sugars, oils and starch, which can be used for
the production of heat or other forms of energy by
being oxidised or burned. The living substance of
the organic cell may be called bioplasm : we do not
know what this is, but when it is killed we recognise
it as protoplasm. The bioplasm of the cells is
continually dying or wasting away and it must be
renewed ; under its control the hard parts of the

body are being formed and the reserve food substances
are being oxidised, or are being decomposed, to
supply heat or energy to the organism. Now the
living cell must be supplied with food to compensate
for its waste ; to form new structures or tissues, or
new individuals during reproduction ; or to form the
skeletal structures ; or to be oxidised for heat and
energy production. The substances which are brought
to the cells for these purposes are the foods, and
their incorporation in the cell substance is the process
of assimilation.

We recognise in the tissues of an animal substances
called proteids, fats and carbohydrates. The fats are
relatively few in number, and when not incorporated
with the proteids are fairly simple in composition.
So are the carbohydrates, and we know the chemical
structure of most of these substances. But the
proteids are exceedingly complex, more so than any
other chemicals known to us, and in spite of long-
continued research we do not know what is their
constitution. But we do know that the proteids of
every species of organism must be different from
those of any other species, and it is possible that this
holds true even for individuals. The facts of heredity
can hardly be explained in any other way, and indeed
there is a certain amount of experimental proof in
support of this assertion. We also know that the
proteids can be split up, or decomposed in various

ways, and that the products of these decompositions
are always substances called amino-acids, which
though very much simpler in composition than the
proteids are still very complex chemical compounds.
We also know that the amino-acids can be re-
constituted to form proteids, but this reconstitution
can only be effected by the agency of a living cell,
and all attempts to make this synthesis outside the
living substance have hitherto failed.

The food-stuffs of an animal are the proteids, fats,
and carbohydrates of some other animal or plant and
they are always different from those of its own body,
and cannot be directly utilised by it as food. Before
these substances can be assimilated they must be
broken down into their constituents and then the
latter must be resynthesised to form the specific
proteids, fats, or carbohydrates of the animal. This
decomposition and recomposition make up the
processes of digestion in the body of an animal. In
our own case the proteids may be those of lean
meat, of white of egg, of cheese, of the gluten of
wheat bread, milk-albumen, etc. The carbohydrates
may be the starches or sugars of various grains, and
the fats are also usually different from those of our
own body.

Digestion of the food is carried out by substances
called ferments, or enzymes, elaborated by the
digestive glands. The proteolytic enzymes are those

which act on proteids ; they are the pepsin of the gastric juice, the trypsin from the pancreas, and the erepsin and enterokinase secreted by the wall of the intestine. They decompose the proteids first of all to peptone and then to amino-acids, but what proportion of the proteid taken into the intestine is broken down so profoundly as the stage of amino-acids is still uncertain. The lipolytic enzymes are those which split up the fats into fatty acids and glycerine : the principal one is the lipase from the secretion of the pancreas. The amylolytic enzymes are those which resolve the starches into sugar or rearrange the sugars into forms which are capable of assimilation. They are the ptyalin of the saliva, and the amylase from the pancreatic juice.

Thus the proteids are converted into amino-acids, the fats to fatty acids and glycerine, and the starches to sugars—all this by the activity of the enzymes of the alimentary canal. These products of digestion —the food-stuffs we may call them—are soluble in water and they are absorbed by the cells lining the wall of the intestine. But these cells also contain the enzymes and the action of the latter is reversible, that is they can not only split up the proteids to amino-acids but they can recombine these substances back again into proteid. So in the process of digestion the amino-acids are reconverted into proteid, the fatty acids and glycerine are reconverted to fats,

and the sugars either pass directly into the blood
stream without further change or they are transformed
into compounds which are capable of assimilation.
But—and this is the aim of the whole series of
digestive actions—the reconverted proteid and fats
are different from those which were taken into the
alimentary canal and are now similar to those which
make up the tissues of the animal. They are the
true foods, and as such they are taken into the blood
stream and are carried over the body to be assimilated
by the cells of the tissues.

The nutritive processes of plants are profoundly
different from those of animals. The proximate food-
stuffs of the plant organism (or the raw materials of
the food-stuffs, as they are sometimes called) are
salts of nitric acid or nitrous acid, salts of ammonia,
and carbonic acid. These are exceedingly simple
substances, and though the plant is able to make use
of more complex nitrogenous stuffs than the salts of
nitric acid or ammonia, yet the latter are a sufficient
source of food. The mineral salts are taken up from
solution in the water of the soil by the root hairs, and
the carbonic acid is taken in from the atmosphere
through the stomata of the leaves. If we are dealing
with a marine plant these substances are simply
absorbed from solution in the sea-water over the
entire surface of the organism. The carbonic acid
and water taken into the plant tissues are combined

to form starch, a synthesis which can now be imitated in the laboratory by agencies available to chemists apart from the activity of the living cell. The latter effects the combination by making use of the energy of sunlight, and it appears that a very small proportion of the total light falling on the surface of the plant may be sufficient for this purpose, nevertheless this fraction is quite essential. In spite of the most careful and protracted investigation the details of this process of the photo-synthesis of starch are imperfectly known; but carbohydrate in the form of starch is elaborated by the plant organism from inorganic material, and is immediately converted into soluble sugar which is the carbohydrate food of the plant.

The simple compounds of nitric acid and ammonia are taken into the tissues of the plant and are then combined with the carbohydrate to form amino-acids, and these food-stuffs are further elaborated into the form of proteid. The sugars and proteids pass into circulation in the vascular tissues of the plant and are assimilated by the cells—either stored up in the form of proteid, oils, or starch, or oxidised to supply energy.

We can exhibit this contrast between the nutritive processes of the plant and animal organism as follows.

The proximate food-stuffs are:

In the animal
- Proteid, splitting to form amino-acids say, leucine, amino-iso-butyl-acetic acid,
  $$\begin{matrix} CH_3 \\ CH_3 \end{matrix}\Big\rangle CH \cdot CH_2CH\Big\langle \begin{matrix} NH_2 \\ COOH \end{matrix}$$
- Fat, say olein ... ... $C_3H_5(OC_{17}H_{33}CO)_3$
- Cane sugar ... ... $C_{12}H_{22}O_{11}$

In the plant
- Nitrate, say sodium nitrate ... $NaNO_3$
- Carbon dioxide ... ... CO_2
- Water ... ... OH_2

The proximate food-stuffs, the amino-acids, the fats, and carbohydrates are the same in each case. The foods are also the same, that is they are proteids, fats, and soluble carbohydrates. The process of assimilation, that is the incorporation of these foods by the bioplasm of the cells is also the same. The difference is in the preparation of the proximate food-stuffs, for in order to obtain these the animal must break down material which it can only obtain from the tissues of other animals or plants—materials which are highly complex in structure, and it must then synthesise these to form its food-stuffs. The plant builds up its food-stuffs from inorganic material possessing almost the simplest chemical structure known to us. The difference is expressed by saying that the functions of the plant organism are in the main synthetic, or constructive ones, while those of the animal are, in the main, analytic, or destructive ones. This is true generally but we must not forget that many of the processes of the animal are also synthetic.

Now there are few conceivable forms of structure that have not been evolved among animals and plants, and we may assume that there are also few possibilities of nutrition that have not also been evolved. We have to consider, then, what other modes of nutrition besides the purely plant one (holophytic) and the purely animal one (holozoic) exist among marine organisms. There is obviously a combination of the two, and this occurs among some molluscs, some polyzoa, at least one starfish, many coelenterates, including most corals, many worms, possibly in some sponges, and in a great number of protozoa. In all these cases the animal which exhibits the plant-like mode of nutrition is coloured green, and this is due to the presence of chlorophyll—the colouring matter of plants, by reason of which they are able to intercept the energy of sunlight and make use of it in the process of photo-synthesis. The best known case of this kind is that of the worm Convoluta, a full account of which is given in the volume *Plant-Animals* in this series of Manuals. Convoluta is green because of the presence in its tissues of cells containing chlorophyll; but it does not necessarily begin life as a green animal, for the green cells are the degenerate remains of algae which have infected the tissues of

7—2

the worm almost as soon as it was born. It is quite
possible to rear the worm apart from the algae as a
colourless animal, and it is also possible to rear the
algae apart from the worm. When it is young the
worm feeds by ingesting diatoms, etc., but after a
time it ceases to take in any solid food. It obtains
its carbohydrate from the activity of the green cells,
which are able to synthesise starch from the carbonic
acid absorbed from solution in the sea-water, and
this starch is converted into sugar and is assimilated
by the other tissues of the worm. The alga obtains
its nitrogenous food from the products of excretion
of the worm, and for a time the association of the
plant and animal in one compound organism is of
advantage to both partners. But since the animal
has lost the power of feeding it becomes unable to
supply itself with proteid food, although it receives
abundance of carbohydrate food from the photo-
synthetic activity of the alga. Towards the end of
its life it therefore begins to digest the green cells in
order to obtain proteid food, and thus by cutting off
its supplies of carbohydrate it dies of starvation.
The earlier stage of the association of the two animals
is one of symbiosis, and it is of advantage to either
partner, but the later stage is one of parasitism on
the part of the worm.

In this and some other cases the association
together of a plant and animal is an obligatory one,

neither organism being able to exist without the other. This, however, is not always the rule, and it is probable that the association may be one which can be lost and regained : that is, the animal may be able to live well enough in some circumstances without the aid of the algal cells, while it is certain that the latter may be able to live as independent organisms in the water, procuring their own proteid food as well as their carbohydrate. Alcyonaria, which are colonial animals, are often coloured green in tropical seas, but they are colourless in British waters. Corals in warm seas are often also coloured because they contain associated algal cells, and some species of the protozoan Noctiluca which are colourless in British seas are green in the tropical waters. We have seen that the tropical seas are poor in plankton, and it is probably the difficulty of procuring sufficient carbohydrate food that has induced the association of chlorophyll-containing algae with the animal organism, for whenever this partnership is set up the difficulty disappears because of the abundance of carbohydrate resulting from the photo-synthetic activity of the algae.

In all these cases which we have considered the association of the two organisms establishes a combination of the two modes of nutrition—the holophytic and holozoic—and this is termed heterotrophic. There is no doubt in such cases that we have to deal

with plant-animals, compound forms brought together
by mutual necessity. But there are also the cases of
many planktonic unicellular organisms, such as the
peridinians, and some others which are more difficult
to understand; for although these organisms possess
chlorophyll corpuscles in their cells so that their
mode of nutrition is truly holophytic, yet their general
characters may suggest those of the animal rather
than the plant. The difficulty, however, is not one
that need concern us, for it may be the case that many
groups of the protozoa have evolved from chlorophyll-
containing organisms which we need not regard as
either plants or animals; and that some of them
while developing along the animal line have still
retained their plant-like mode of nutrition.

Parasitism, to which we have already referred, is
essentially a mode of nutrition in which the process
of digestion is either eliminated entirely, or is greatly
abbreviated. In this condition two animals or two
plants are associated together, or an animal may be
associated with a plant. The association is often said
to be an obligatory one in that one of the associates,
the parasite, must live within the body, or on the body
of another, the host; although in few cases is the
actual contingency of the associates strictly necessary
for the life of the parasite. The parasitic habit is
very common among marine animals and we are able
to distinguish between two main groups of animals

pursuing this mode of life: (1) animals which live attached to the outer surfaces of their host—the skin, the gills, the branchial cavities, and the cavities of the nose or mouth. Examples of this class of parasites are the Copepods (fish-lice), which are attached to the external surfaces of many species of fishes; some leeches; and some Trematodes (flukes). (2) Internal parasites which live in the alimentary canals of their hosts, either holding on by means of suckers or hooks, or lying quite free in these cavities (tapeworms, internal trematodes, and thread-worms); or which live in the blood-stream or lymph channels, or in the peritoneal or serous cavities; or embedded in the substance of the muscles or other tissues (trematodes, thread-worms, larvae of tapeworms, trypanosomes, etc.). The external parasites usually pass through a life-history which includes a free-swimming stage in which the organism lives among the plankton as an independent animal, and then at a certain phase, coming in contact accidentally with its host, it attaches itself thereto and undergoes metamorphosis to its sexually mature form. Some of the internal parasites also pass through a free-swimming stage in the course of which they are taken into the alimentary canal, or into some other part of the body of an animal which acts as their larval host. They live in the tissues of this host and do not undergo development beyond a certain stage, and finally degenerate unless the larval host is eaten by another animal which acts as the adult host. When this occurs the parasite undergoes full development and attains sexual maturity, and its eggs pass out of the body of the host and infect further larval hosts. In other cases still the parasite never lives in the open except as an egg which is eaten by the larval host, and this in turn is eaten by the adult host.

The parasitic habit is advantageous to the parasite but is detrimental to the host, since substances may be excreted by the parasite which may be to some extent poisonous. The association is also harmful to the host since the parasite uses up the food-stuffs of the latter. A typical parasite lives in the alimentary canal of another animal and it is bathed in the food matter which is undergoing the process of digestion. It may or may not possess an alimentary canal and mouth: no tapeworm does but most trematodes do. If an alimentary canal is absent the parasite simply absorbs the soluble food matter of the host through its skin; if it is present the food matter may be absorbed through the skin or taken into the mouth. In whatever way the food-stuffs are absorbed they always consist of soluble peptones or amino-acids, of soluble fatty acids and glycerine, and of soluble sugars; and all these substances have been prepared by the digestive enzymes of the host. This is the complete elimination of the process of digestion and all the parasite has to do is to recombine the food-stuffs into the specific proteids, fats, and carbohy-drates of its own tissues. The process of digestion is abbreviated in the case of those parasites which live on the external surfaces of their hosts, for there the food consists of mucus which is probably easier to digest than the proteid of a captured animal. In the most profound parasitic conditions, as when such a form as a trypanosome inhabits the blood-stream or the cerebro-spinal fluid; or when a blood-sucking parasite lives on the external surface of its host with its suctorial mouth embedded in one of the larger blood-vessels, it simply absorbs the ready-made proteids of the host, possibly in the same way as the cells of the latter also absorb them. Essentially then the parasitic mode of nutrition is one in which the digestion of food-stuff is performed outside the body of the organism and by the agency of some other organism. It may be called the saprophytic mode of nutrition when we speak of a plant parasite, and the saprozoic mode when we speak of an animal parasite.

Now let us consider parasitism in the open—that is a parasitic mode of life practised by an animal which does not live in the cavities, or on the body of another animal. It seems paradoxical so to speak, but we must remember our definition of the habit—that it consists in essence of the utilisation of already digested food; and further that it is possible to rear many parasites outside the body of other animals in artificially prepared juices or jellies; thus many species of bacteria may live either in the body of a host or in the sea-water or mud, but their mode of nutrition is the same in each case.

Are there animals which live in the open and which do not ingest solid food in the shape of the bodies of other animals and plants, but live by ingesting matter which is dissolved in the sea-water, in short are there marine animals which do not eat visible food? Now it is not at all remarkable that zoologists, whose attention is usually concentrated on the study of form rather than function—who look on an organism not so much as 'something happening,' but as a structure of skeleton, muscles, alimentary canal, glands, nervous system, etc.— should have regarded such a mode of nutrition as improbable. Why should an animal possess an alimentary canal and glands if these structures are not for the purpose of the digestion of food? Yet the study of parasitism and its modifications should have led us to postulate the existence of a widespread saprozoic mode of nutrition, at all events among the lower marine animals. We might not have expected to find it among the terrestrial animals but where an organism lives continually bathed in a liquid medium, the absorption of food matter through its outer surface—apart altogether from an alimentary canal—should always be regarded as a possibility of nutrition.

If we assume that marine animals living in the open nourish themselves exclusively by capturing and digesting solid food we immediately get into difficulties.   For it is often difficult to convince oneself, by examination of the contents of the alimentary canal, that sufficient food organisms for the apparent needs of an animal have been ingested. We hardly find anything in the intestine of a lugworm except sand which does not differ from that in which the animal lives.   It burrows by eating the sand in the same way as an earthworm burrows by eating the soil, and we suppose that its food is contained in the sand that passes through its body. But if its food consists of microscopic organisms there are very few such in the sand beneath the superficial layers.   Many organisms have indeed been identified from the contents of the intestine of the lugworm but it is always possible to regard these merely as residues which are contained in the sand, and it does not appear to be probable that the worm can obtain enough food-stuff in this shape from the sand which it ingests.   Just the same difficulty meets us in the cases of cockles and mussels which live on the sea-bottom and which have been supposed to obtain their food from the plankton contained in the water which passes through their shell cavities.   If we

examine the contents of the alimentary canal of
these animals we nearly always find that it contains
hardly anything but fine sand and mud particles.
There are never many food organisms, such as diatoms.
It has been said that these animals may feed on such
organisms as infusoria which possess no shells and
which could, therefore, hardly be identified in the
contents of the intestine, but we do not find many
such if we examine the water in which the molluscs
live. Sometimes indeed we do find that the intestine
contains large quantities of green matter which
probably consists of algal spores, but this is exceptional. If, however, we remember that it is
essential for the good condition of such molluscs as
these that they should live in a part of the sea
where there is plenty of fresh water flowing down
from rivers, then we receive the suggestion that
probably the food may be in solution, for the water
of rivers flowing into the sea contains more of
dissolved organic compounds than does sea-water.

Then it can be determined in many cases that
the quantity of planktonic organisms contained in
the plankton is too small to account for the observed
metabolism of an animal. A mussel, for instance,
may very easily grow from two inches in length to
two and a half inches in the course of a single
summer, and it was easy to show in one such series
of cases that this growth represented an average

gain in dry organic substance of 713 milligrams, say 356 milligrams of proteid, since analyses of mussel flesh give about half the dry weight as consisting of this constituent. Now let us suppose that all this proteid came from the plankton captured by the shell-fish. A good estimate of the amount of plankton contained in northern waters is that 1000 litres of water contain, on the average, 168 milligrams of dry organic substance, and since we know the composition of various forms of plankton we can easily calculate the amount of proteid contained in the water : it is 25·2 milligrams. If the solid matter filtered from the water by the mussel were as rich as this—and it certainly is not, for the greater part of it consists of fine mud, and if the animal captured all the solid matter contained in the water sucked into its shell—and again it certainly does not—then it would have to filter 14,200 litres of sea-water in order to obtain this amount of proteid. But even then it must have digested and assimilated all the food matter taken into its intestine—and the digestive machinery of no animal is as efficient as this. Above all it must have built up all the assimilated proteid into the form of tissue, and it certainly could only have so disposed of a small fraction of it for the largest part by far must have been oxidised in the production of energy. This example gives us an idea of the difficulties we saddle ourselves with when

we try to show that such an animal as we have been
discussing nourishes itself in the same way as a
mammal does.　Again a sponge causes a current of
water to flow through the system of cavities per-
meating its body, and it nourishes itself by removing
the food matter contained in this water of circulation.
If we estimate the quantity of carbonic acid excreted
by the sponge in the course of an hour, say, we can
find what was the quantity of carbon contained
in the parts of its body which were oxidised during
that time—that is to say we find what was the
quantity of carbon contained in the food which it
must have taken in order to keep itself from losing
weight.　Now we know what quantity of carbon is
contained in the water in the form of plankton and
it is therefore easy to calculate what quantity of sea-
water must be filtered by the sponge in order to get
enough food, on the assumption that all its food is
contained in the plankton.　The volume of water so
found is ridiculously large—far too large to make it
possible for the sponge to have filtered it in the time
required.

We can show also, by actual experiment, that
many marine animals, even fishes, can be kept for
many months in water which is filtered so that it
does not contain any plankton.　The fish or other
animal may live in good health for months in such
circumstances provided the water is kept cool, and is

frequently changed ; and we therefore say that it must receive abundant supplies of oxygen for the purposes of respiration. But the whole object of respiration is the oxidation of the substance of the body in order to yield the animal the energy it requires, and if respiration continues there must be a continual wastage of the tissues or reserve substances of the body. This waste is either replenished by the assimilation of food, or the animal must lose in weight, having used up its own body substance. Now the loss of tissue due to respiration can be calculated from the amount of carbonic acid excreted and it may be greater than can be accounted for by the loss of weight of the animal. In such a case as this the animal must have nourished itself in the manner of a parasite—saprozoically, by the absorption of dissolved food matters.

Why then should an alimentary canal have been evolved among marine animals if they can nourish themselves by absorbing food from solution in the sea-water ? In attempting to answer this question we will assume that primitive marine organisms were microscopic in size and that they nourished themselves by obtaining their food-stuff from solution in the water, just as a peridinian does at present obtain its nitrogenous food-stuff. It gets its carbon food by building it up from the carbonic acid which it also absorbs from solution, but it cannot do this unless it also obtains nitrogenous food, and it must get this from the amino-acids or analogous substances which it finds in solution in the sea. Myriads of marine unicellular animals so obtain their proteid food-stuff from soluble substances and they can do so easily because they are small. If they are very small their surface is large compared with their bulk, and they absorb food-stuff in proportion to their surface. Now let the organism increase greatly in size and its surface must diminish relatively to its volume; for the surface with increasing size is proportional to the square of the radius, while the volume is proportional to the cube of the radius. As the organism grows in size it must therefore find increasing difficulty in absorbing food-stuff from solution in the sea-water; for its requirements are proportional to its volume while its powers of obtaining food are proportional only to its surface. Further, the absorptive surface must become reduced as the animal increases in size, since part of it at least may become thickened to form an integument, so as to afford protection from mechanical injury, and the powers of absorption must therefore be reduced.

The surface must therefore be increased during the evolution of increasing size, and this is effected by the infolding of a part of the body-wall so as to create an internal cavity, the wall of which may remain thin because it is protected by its situation. Such an internal cavity we find in Hydra and in coelenterates generally. Let the cavity open at both ends —a mouth and an anus—and we have an alimentary canal. Even then the surface may be insufficient, for as the animal increases in size its movements through the water become slower when compared with its size ; but if it should begin to swallow water and pass this through its alimentary canal the absorptive power would still further be increased. If it should still be insufficient, a further increase in absorptive surface may be attained by the outfolding of the body-wall so as to form tentacles, gills, plumes, etc., structures which have a large surface compared with their volume, and the walls of which may remain thin if they are protected from mechanical injury in some way—as in the case of the gills of a fish. Now let the animal still further increase in size and two new devices may come into operation: (1) a circulation of blood or other internal fluid is initiated, and this is most abundant in the outfoldings of the body-walls and round the wall of the internal cavity. Since the absorbing fluid is thus continually being changed, obviously more food-stuff can be taken in. (2) a circulation of water is established in relation to the gills, or outfoldings, by means of cilia, or other mechanisms, and thus the sea-water is caused to flow more rapidly over the absorptive surface. Since the source of soluble food-stuff is thus continually being changed more of the latter can be taken up than if the water were at rest.

Thus we can account for the evolution of an alimentary canal, mouth and anus, gills, blood circulation, and respiratory movements and mechanism, solely by reference to the necessity for the increased absorption of dissolved food-stuff, and without reference at all to the digestion of solid food matter ; and it is the increase in size of organisms which has led to the necessity for this development.

But nevertheless there is a digestive mechanism in many animals, and we have also to account for the development of this even if the absorption of soluble food remains a factor in their nutritive process. Now the key to this is the establishment of an internal cavity. Some insectivorous plants have such an internal cavity, evolved by leaf modifications. It happens that small animals enter these cavities and they are utilised as sources of food-stuffs by the plants, but they are not necessarily digested. Utricularia does not digest entrapped flies, but the latter die in the bladders and undergo bacterial decomposition. Drosera does possess a digestive fluid but this only carries on the process as far as the formation of peptones. Nepenthes forms a true proteolytic enzyme. We can easily conceive of the evolution of such a digestive fluid in the internal cavities evolved for the purpose of increasing the absorptive surface of animals. Small particles taken into them would be ingested by the cells lining their walls just as small particles are ingested by the amoeboid cells of the alimentary canals of many worms, and just as an amoeba ingests its prey. But if the food particles were too big to be ingested in this way? Even then we need not postulate the existence of a digestive fluid, for autolysis of the food organism would occur when it died. Every cell of the body of an animal contains proteolytic, lipolytic, and amylolytic enzymes, and when a tissue dies self-digestion, or autolysis, begins. It is by autolysis that the exudation producing consolidation of the lung disappears in cases of recovery from lobar pneumonia. Or bacterial decomposition of the body of the entrapped food organism would occur. The same result would be attained either by autolysis, bacterial decomposition, or by digestion—that is the resolution of the proteids to the stage of amino-acids, that of the fats to fatty acids, and that of the carbohydrates to soluble sugars. All these substances would then be absorbed by the walls of the internal cavity in which they lay.

We may however reasonably suppose that a cell in contact with an eatable organism will be stimulated

to produce enzymes in greater quantity when it is
in contact with a trace of the decomposition products
of a food organism ; that is a signalling substance,
or hormone, would be produced just as it is in the
mammalian intestine, and this would stimulate the
cells to elaborate the enzymes. But if the food
organisms were too big for the cells to ingest then
we may suppose that their contained enzymes would
diffuse out into the cavity, that is to say, a 'secretion' would take place. This is indeed the manner
of digestion in many invertebrates. Molluscs, for
example, possess a 'digestive gland,' 'hepato-
pancreas' or 'liver' as it is called, and there are
accounts of the 'secretion' from this organ and of
the reactions of the fluid said to be so produced.
There is really no evidence of the production of a
secretion in the sense in which we employ the word
in the physiology of the mammal. The tests made
involve the pulping of the gland and the examination of the product, and the latter certainly contains
the intra-cellular enzymes of the cells of the gland.
The latter is only an extension of the cavity of the
intestine of the mollusc.

There is of course no reason why the preparation
of the food by the digestion of captured organisms,
and its preparation by the absorption of dissolved
food-stuff should not proceed simultaneously in an

animal, just as the preparation of food by the digestion of the body of a captured insect, and its preparation by the process of photo-synthesis go on simultaneously in insectivorous plants.

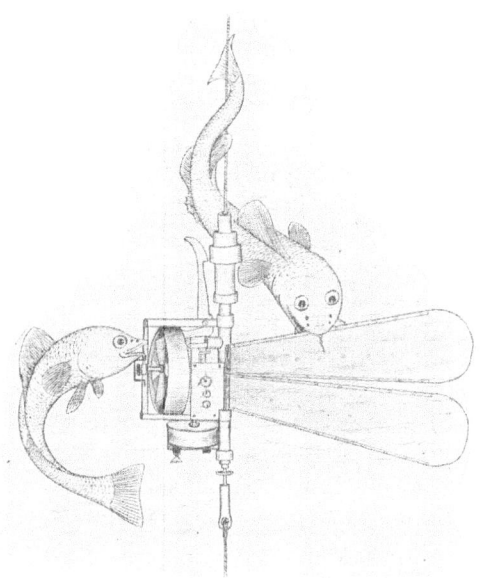

The Ekman Current-Meter.

CHAPTER V

THE SOURCES OF FOOD

WE have not yet considered all the known modes of nutrition for it will be more convenient to discuss some of them when dealing with the question of the circulation of food-stuff in the sea. We shall not be able to understand the meaning of the more obscure modes of nutrition unless we have clear ideas as to the assimilation of food. All organisms, whatever be their mode of life, have this in common, that their bioplasm continually wastes away by its own activity and must be renewed ; and further, to obtain energy all must oxidise or transform some substance which the bioplasm associates with itself. The substance which they use in order to renew their bioplasm is proteid, and that which they use to obtain energy is called the respirable material. The latter may be one of several substances.

In green plants and animals the respirable material is some form of fat or carbohydrate, and both forms of organisms oxidise this to carbonic acid and water, which are then excreted into the surrounding air or water. In oxidising this carbonaceous food heat, or some other form of energy, is produced. The carbonaceous food consists of carbon and hydrogen, or of these elements and oxygen as well. All forms of carbohydrate and fat are built up of chains of compact little groups of atoms ; thus grape sugar has the following composition :

$$COH.(CH.OH)_4.CH_2.OH.$$

The proteid is much more complex than any other substance known to us. It is built up from amino-acids, which are chains or rings of little groups of carbon and hydrogen atoms as in the case of the carbohydrates, but in addition to these the amino-acid contains a group of nitrogen and hydrogen atoms—the amino group, NH_2—and this is associated in a certain position with another group, the carboxyl one, $COOH$. Amino-acids are piled on amino-acids to form the giant proteid molecule, and this differs from the other food-stuffs, not only in its greater complexity, but also in the fact that it contains nitrogen. In order to manufacture their proteid food all organisms require nitrogenous food-stuff.

Two sharply contrasted modes of nutrition have forced themselves on our attention—the holophytic and the holozoic. In the former the carbonaceous food-stuff is the carbonic acid of the atmosphere, and the nitrogenous food-stuff is some salt of nitric acid or ammonia. The respirable material is the carbohydrate built up from the carbonic acid and water, and this is then combined with the amino-acid formed from it and the nitrogenous food-stuff to form the proteid. These processes are chemical transformations which we can just imitate in the laboratory. In holozoic nutrition the carbonaceous food-stuff is fat and carbohydrate, and these are converted in the tissues into sugars and fats which are the respirable materials ; and the nitrogenous food-stuff is proteid which is converted into the specific proteid of the animal after having been digested. The mode of nutrition of the plant-animals is a combination of these two primary modes ; and that of the parasites and other saprozoic organisms is simply the holozoic mode abbreviated in some way.

All animals, whether they live in the open, or are parasites, or whatever be their manner of feeding, must nourish themselves by ingesting the dead or living bodies of other animals or plants, or proteid or amino-acid derived from these sources; and they can only obtain their carbonaceous and nitrogenous food-stuffs in this way, and not from inorganic materials. Animals must eat other animals or plants. Let us think of an ocean containing only animals and we see that the larger must eat the smaller ones in order to live, until in the end there would only be one animal left and it would die of starvation. The animals are consumers, and their life-process may be regarded as that of a clock which is always running down and cannot be wound up except from the outside of itself.

The plants are producers for they can convert inorganic into organic living substance. Carbonic acid, salts of ammonia, and nitric acid exist in nature and can come into existence apart from the agency of life, and these very simple substances can be synthesised to form proteid and carbohydrate. All living substance therefore arises from the carbonic acid of the atmosphere or the sea-water, and from simple mineral nitrogen compounds, mainly by means of the agency of plants. The pastures of the land are the main sources of terrestrial animal life, and the diatoms, peridinians and other vegetable planktonic organisms are the pastures of the sea.

All animal life is destructive and its continuance implies the degradation of chemical compounds. Proteid, carbohydrate and fat are compounds possessing high potential energy, that is they can be oxidised or burned, as coal is burned, giving a large amount of heat or energy; or they can be decomposed, as an explosive is decomposed, also yielding heat and energy. When taken into the tissues they are oxidised or decomposed and energy is obtained, and the carbon and hydrogen which they contain leave the body in the form of carbonic acid and water, mainly from the lungs and kidneys. The proteid is also oxidised or decomposed and most of its carbon and hydrogen also leave the body in the form of carbonic acid and water, but the nitrogen part of the molecule is not completely degraded. Carbonic acid and water are compounds which are fully oxidised and can no longer be used to yield energy but the product of the nitrogenous metabolism is either urea, or uric acid, or some other analogous substance which is not fully oxidised and can still supply energy. With the exception of these nitrogen compounds the clock has run down.

When an animal or plant dies the substance of its body becomes degraded so that the ultimate products of decomposition are water, carbonic acid, nitric acid, sulphuretted hydrogen and a few other substances. This resolution of the organic body is effected by the agency of a number of species of bacteria. The proteids are attacked by the putrefactive micro-organisms which first of all reduce them to the form of amino-acids just as in the case of digestion. The fats and carbohydrates are attacked by the fermentative micro-organisms which reduce them ultimately to the form of carbonic acid and water. Some of the microbes of fermentation and putrefaction act with the assistance of oxygen (aerobic bacteria), while

others (the anaerobic bacteria) can act in the absence
of oxygen except that which is contained in the substance which they are attacking. When putrefaction
takes place in the absence of oxygen the breakdown
is much more complex and evil-smelling compounds
are produced: in the presence of oxygen the process
of putrefaction is more rapid. The bacteria of putrefaction and fermentation have a mode of nutrition
which is very similar to that of the saprozoic animals,
but they are living among abundance of food and
their metabolism is very wasteful. That is to say
they break down a large quantity of proteid or
carbohydrate in order to obtain a small quantity of
energy; and the total amount of bacterial life in a
fermenting or putrefying mixture is far less than
could be sustained if the latter were used to the
greatest advantage. Now we cannot say with con-
fidence that the bacteria are either plants or animals
so we hedge, and speak of their mode of nutrition
as metatrophic.

In the long run the products of bacterial action
on organic matter are mainly ammonia, carbonic acid,
water, sulphuretted and phosphoretted hydrogen.
Now the urea excreted by animals undergoes change
to ammonia and carbonic acid, either by means of an
enzyme, or by bacterial action, so that the excretory
products formed by an animal during its life are in
the long run the same as those which are formed

after death by the decomposition of the substance of its body: that is, they are carbonic acid, water, ammonia with traces of sulphuretted hydrogen and phosphoretted hydrogen. The nitrogen which was in the proteid is not however completely degraded for it can be oxidised and can still yield energy. But we do not find ammonia in nature except where it is being produced, as in the emanations from volcanoes, or in decomposing matter, such as farmyard manure, for example. It is never stored in nature, and the only natural nitrogen compounds are nitrates, chiefly Chili saltpetre (sodium nitrate). The conversion of ammonia to nitric acid is a process of oxidation and it is accomplished by the agency of bacteria.

These nitrifying bacteria introduce us to another mode of nutrition of extraordinary interest. They exist wherever there is plenty of oxygen and ammonia. There are various species—the same apparently in all parts of the world and sea where they have been investigated. One species oxidises ammonia (NH_3) to nitrous acid (HNO_2) and another species further oxidises the latter compound to nitric acid (HNO_3). They do not require organic matter as a source of food, and are indeed inhibited by the presence of such; and like green plants they can form proteid and carbohydrate from simple inorganic compounds of nitrogen, and from carbonic acid and water. But a green plant can only do this in the presence of sunlight, which is employed by the chlorophyll it contains; while the nitrifying bacteria can do all that a green plant can do, in the absence of chlorophyll, and in the dark. Of all organisms theirs is the simplest and most profound mode of nutrition. We have to invent a new term to describe it and we call it prototrophic.

There remains the sulphuretted hydrogen and the phosphoretted hydrogen which are produced during decomposition or during excretion. These also are compounds which are not entirely degraded and still possess energy in that they can be oxidised. The phosphoretted hydrogen is at once oxidised in the air to phosphoric acid, and the sulphuretted hydrogen may also be directly oxidised. But wherever in the sea, or in fresh water, there is sufficient of this gas there are also red or colourless sulphur bacteria, and both of these forms of microbes are able to use the gas, which is poisonous to the higher organisms, as a source of food-stuff. A mere trace of carbonaceous food in the form of some simple compound, such as formic acid for instance, and a trace of nitrogenous substance in the form of ammonia are sufficient for the renewal of their bioplasm, and in place of the soluble sugar, which is assimilated as respirable material by the higher organisms, these bacteria can assimilate the sulphuretted hydrogen, oxidising its hydrogen to water, and setting free the sulphur in their cells. This process it will be seen is analogous to the assimilation of carbon dioxide by the green plant and the storage of starch or sugar in the cells. The sulphur is then oxidised to produce energy, and ultimately forms sulphuric acid, when it cannot be oxidised further. The sulphur bacteria are, like the nitrifying bacteria and the green plant, prototrophic in their mode of nutrition, and can form proteid from inorganic materials, but they can accomplish this in the absence of chlorophyll and sunlight.

Thus we have accounted for the degradation products of the animal body whether these result from the waste during life or the decomposition after death. The nitrogen finally becomes nitric acid, the hydrogen water, the carbon carbonic acid, the sulphur sulphuric acid, and the phosphorus phosphoric acid. The trace of iron also present in the body becomes altered by the action of the iron bacteria to ferric oxide. Plant tissues may also undergo decomposition in such a way as to form masses of nearly pure carbon, as in the formation of coal and anthracite, and there is now evidence that even this amorphous carbon may be attacked by certain bacteria which can oxidise it to carbonic acid with the production of heat. Thus all the products of destructive metabolism (katabolism) tend to pass into the state of fully oxidised compounds so that they can no longer be made use of by the animal as sources of energy.

Their elements are the same as those which make up the greater mass of the animal body, but they have passed out of circulation so far as the nutrition of the latter is concerned.

But the degradation products of the animal body are the sources of food-stuff for the plant organism, and just where the destructive processes of the animal end, there the constructive processes of the plant begin. The simple compounds, carbonic acid, nitrate or ammonia, are utilised by the algae, the diatoms, the peridinians and other protozoa which possess chlorophyll, and by all the higher animals which also possess green cells in their tissues. These substances are those on which all the life in the sea depends for they are employed in the constructive work of the plants. They are the raw materials for the production of living substance—the ultimate food-stuffs of the sea.

It is important then that we should have a knowledge of the proportions in which these substances exist in sea-water, and it must be confessed that our knowledge on this point is not so full as is desirable. Now it is to be noted first of all that not all the proteid and carbohydrate and fat of animal or plant tissues suffer, at once, the profound degradation which we have mentioned above. Many organic carbon and nitrogen compounds enter the sea in the water of the rivers, or pass into solution from organisms living and dying in the sea. These carbon compounds are generally similar to those found in the soil and known as humus—they consist of fatty acids and soluble carbohydrates ; the nitrogen compounds are such as are described by the water analysts as 'albuminoid ammonia,' that is amino-acids and analogous compounds. They would finally be reduced to their simplest terms by the action of bacteria, but before they become so oxidised part of them is made use of as food-stuff by the saprozoic animals and the saprophytic plants.

Trustworthy analyses made from the water of the North Atlantic show that these substances are present in the following proportions :

Carbon compounds other than carbonates dissolved in sea-water—9·2 milligrams per litre (9·2 parts per million).

Nitrogen compounds other than ammonia or nitrate dissolved in sea-water—0·126 milligram per litre (0·126 part per million).

The amount of these substances at the disposal of saprozoic organisms is therefore very small, but it is greater than the amount of proteid or carbohydrate contained in similar volumes of sea-water in the form of plankton. Thus some quite trustworthy estimations of the food value of the plankton give the following results (northern seas):

Carbon present in the form of organisms—0·083 milligram per litre (0·083 part per million).

Nitrogen present in the form of organisms—0˙008 milligram per litre (0˙008 part per million).

Thus there are apparently about 110 times as much organic carbon, and about 16 times as much organic nitrogen present in solution in sea-water as is contained in the form of the organised substance of plants and animals of the plankton. These substances are capable of assimilation by saprozoic animals and we may take it that there are creatures in the sea which do feed on them. Nevertheless this does not prevent the final destruction of organic carbon and nitrogen compounds, for of all such resulting from the waste of the animal metabolism only a part will be directly utilised in this way by the saprozoic animals, the rest being broken down by bacteria. And the part thus built up into animal tissue is again broken down in metabolism and part of it becomes the prey of the bacteria. Finally all must undergo this fate.

We know a great deal more with reference to the amounts of the ultimate food-stuffs present in the sea. The amount of carbonic acid varies very greatly according to the salinity, the temperature and the depth. It may be stated in round numbers to amount to 50 milligrams per litre of sea-water. There are also a good many facts with regard to the amount of inorganic compounds present in the sea. Analyses by modern methods have been made from samples of water taken from the North Atlantic, the equatorial seas, and the Antarctic Ocean. The following figures apply to northern seas, and each is an average based on from 15 to 30 samples made in each month of each of the years mentioned.

Inorganic nitrogen compounds present per litre of water from the North Sea and Baltic in the years 1904–6. *The figures represent milligrams. The proportions are therefore parts per million of sea-water.*

Month	Nitrogen from ammonia	Nitrogen from nitrites and nitrates	Total inorganic nitrogen
February	0·061	0·149	0·210
May	0·073	0·136	0·209
August	0·065	0·087	0·152
November	0·073	0·088	0·161

We may consider along with these results those of the main voyages of investigation of late years. Combining the results and 'rounding off' the numbers we find that:—

Water from the Antarctic contained about 0·5   part per million
,,   ,,   Equatorial seas   ,,   0·1   ,,   ,,
,,   ,,   North Atlantic   ,,   0·15 to 0·5   ,,   ,,

of inorganic nitrogen compounds.

All these results apply to water taken from the

surface of the sea at some distance from the land. Many analyses have also been made of coastal water and water from the deep, and without quoting these we may remark that the coastal water contains more inorganic nitrogen and more carbonic acid, and also more organic dissolved carbon and nitrogen compounds than does the water at some distance from the land. The amounts of these substances also increase regularly the deeper in the sea we go, and the difference between the surface and the bottom is very like the extreme difference found in the surface waters of the Antarctic and tropics.

Considering the results a little more closely we notice that: (1) The amount of inorganic nitrogen in sea-water is very small, varying from one-tenth to about one-half per million of water; nevertheless upon this minute trace depends all the life of the sea: it is so small because it is continually being absorbed by plant organisms. (2) It is greatest in the Antarctic where the temperature was nearly that of freezing water; much less in the North Atlantic, where the temperature varied from about 5° to 10° C.; and least in the tropics, where the temperature was about 28° C. (3) The proportion of inorganic nitrogen was least in August, when the temperature was highest; and greatest in February, when the temperature was lowest. (4) The proportion was greatest at the bottom of the sea and least at the surface. It is greatest in the oozes at the bottom. It is large near the land.

In considering the distribution of marine animals and plants we noticed that the abundance of food was a factor of importance and we see now that the abundance of food, that is, the abundance of the plants—the producers of animal food—depends on the nitrogenous and other ultimate food-stuffs present in the sea-water. We need only consider the distribution of the nitrogen for this is the substance which rules the production in the sea. It is true that carbonic acid, mineral salts and oxygen are also required, and that lime and silica are necessary for the shells or skeletons of marine organisms. But when a number of indispensable food-stuffs are required and when one of these is present in small proportions, then the production depends on the substance which is present in minimal quantity, just as the strength of a chain depends on that of its weakest link. The silica and phosphates are present in the sea in very small traces, but they are usually more abundant than is necessary when the amount of nitrogen present is considered. The other materials, and oxygen, are present in relatively large amount.

Let us return now to some of the physical questions discussed in former chapters. We have seen that light penetrates to a depth of about 400 fathoms at the most, and probably effective light

disappears far above that level. It is only in the
upper strata of the sea that plants can reproduce
and grow, for only there do they receive the energy
of sunlight necessary for the photo-synthesis of
carbohydrate food-stuff. It does not help them that
practically unlimited quantities of carbon dioxide
can be obtained from the atmosphere, and unlimited
energy from solar radiation, for their production can
only keep pace with the limited amount of nitrogen
compounds present in solution in the water. On the
other hand it is of no avail that more nitrogen is
present in the depths, for in the absence of the power
of photo-synthesis of starch, no proteid can be formed,
and unlimited nitrogenous food-stuff would be of no
use to them. Production goes on only in the upper
well-lighted layers of the sea and the dead bodies of
plants and animals living there fall to the sea-bottom
and accumulate in the oozes, the globigerina and
diatom deposits for instance. Now it is doubtful
how far bacterial life can flourish at the sea-bottom
where the temperature is very low, and some samples
of oozes examined seem to have been sterile in the
media employed for cultures. Some amount of
bacterial action may go on, but what it is exactly
we do not know—research in this direction is urgently
needed. We may however conclude that the processes
of putrefaction proceed much more slowly at the
bottom of the ocean than on the land, or in shallow

waters, or at the surface of the sea ; and that the proteid, carbohydrate, and fat of the bodies of the organisms falling to the bottom may remain for some time not much altered in composition. They thus form the food of the abyssal animals, and the latter must be, to some extent, carrion-eating creatures. In the end, however, the carbon and nitrogen compounds of the bodies of organisms at the sea-bottom will be transformed to products like urea, and, of course, carbonic acid ; and the urea, or similar substance probably passes into the form of ammonia by the action of enzymes. Then these substances must be removed from the sea-bottom, for otherwise the bottom water would be richer in organic matter than it actually is.

This removal is effected by the agency of currents. In the main the system of oceanic circulation consists of a bottom-drift from about the region of the temperate seas to the equator. This cold bottom water rises up to the surface at the equator, and then becoming heated flows away to the north and south. There is a similar bottom-drift from the temperate regions towards the poles, and then this water rises to the surface to take the place of that which flows to the south or north after it has become lighter by freezing. Thus there is a continual streaming of water along the sea-bottom and this water rises to the surface in the equatorial and polar

seas. It contains an excess of organic matter which
it has removed from the abyssal levels of water.
Even in shallow seas, near the land, or at the
junction of warm and cold currents, there are also
these vertical currents from the bottom towards the
surface. Seasonal changes of temperature must also
produce them, for in the winter the surface waters
must cool and sink to the bottom. Wherever there
is a vertical current coming up from the sea-bottom
there should be an increase in the abundance of
plant life, for this vertical current should contain an
excess of ultimate food-stuffs. For the same reason
there should be a greater amount of life in the
shallow water near the land.

We do actually find this abundance of plant life
near the land, and it is pretty certain that local
abundances of plankton are sometimes due to upwelling currents bringing abundant food-stuff from
the bottom. The richness of plant life in the
plankton of the circumpolar seas is also due in part
to the upwelling currents. There are great differences
from year to year in the abundance of fish life in the
Norwegian seas, and also in the time of spawning, and
in the condition of the fish ; and it is remarkable
that it is the years when the temperature of the sea
is lowest which yield the greatest abundance of life.
Now this has been explained by attributing the
lower temperature of those years to a stronger drift

of water from the Arctic seas, and this stronger drift would bring with it a greater supply of food-stuff.

But why do we not get a corresponding abundance of plant life in equatorial seas where the upwelling of bottom water is maximal, and where nitrogen and carbon food-stuff must be transported from the bottom to the surface? As we have seen, the opposite is the case, and the tropical seas are relatively poor in plant life. This poverty might be explained in more than one way, by the temperature hypothesis for instance, which we have already noticed in Chapter III. But if this were true we should still have an abundance of nitrogen compounds produced by the wasteful metabolism of the organisms there, and it is the case that nitrogen compounds are scarcer in equatorial seas than elsewhere in the ocean. A further possible explanation is that these substances are destroyed by bacteria.

Nitrogen compounds *must* be removed from the sea in some way for they are continually being added to it in the water flowing down from the land in the rivers, and it has been estimated that about 39 millions of tons of this element in the form of dissolved inorganic and organic compounds are added to the seas of the world every year. The amount that flows into the North Sea itself is said to be at least about 390 millions of kilograms annually, while the amount of nitrogen in the form of fish and other

economic products that is taken from the North Sea
in the same time is only about 30 millions of kilograms.
The excess must therefore be removed for the sea
does not appear to be getting richer in these
compounds. No deposits containing nitrogen are
formed in the same way as calcareous deposits are
laid down, but it has been suggested that ammonia
is given off from the surface of the sea into the
atmosphere, and that it is then redeposited on the
land in the rainfall. But this is unlikely and it is far
more probable that the excess is removed from the
sea by the agency of bacteria.

Denitrifying bacteria have long been known from
cultivated land, and they have also been discovered
in the sea near to the land. They are organisms
which appear to live under a variety of conditions :
some requiring simple carbonaceous food while
others must receive more complex substances. Some
of them can act in the absence of oxygen while others
require this gas. They obtain their nitrogen food
from nitrates, nitrites and ammonia, and their life-
process with respect to these substances is one of
deoxidation, for they can reduce the nitrate to nitrite,
the nitrite to ammonia, and the latter substance to
free nitrogen. The free nitrogen so formed passes
into solution in the sea-water, but the latter already
contains as much as it can hold, so the nitrogen then
passes back again into the atmosphere. Now it has

been found that the activity of the denitrifying bacteria is greatest at about 25° C., that is approximately the temperature of the tropical seas, and it is practically arrested at freezing point. Therefore if these bacteria are universally distributed over the seas of the world they must be most active in equatorial regions and least active at the bottom of the ocean, and in circumpolar seas. It is possible to account for the destruction of the excess of nitrogen compounds entering the sea by assuming their activity. If they are present in tropical waters we may also account for the poverty of these seas in nitrates and nitrites, and we should have a convincing explanation of the poverty of tropical seas in vegetable plankton.

One is tempted to follow up these hints by a speculation as to past modes of nutrition of marine animals, and as to the kinds of food-stuff which they had at their disposal. No fact of biological chemistry is stranger than this—that nitrogen should be the element which is so intimately associated with the processes of life. For of nearly all chemical elements it is this one which is the most sluggish and inert. Yet its compounds with carbon, hydrogen and oxygen are those which exhibit the greatest degree of complexity known to us, and the reactions of these substances with each other, and with simpler compounds constitute what we recognise as the phenomena of life.

All writers who have considered the question of the circulation of nitrogenous food-stuff in nature have pointed out the remarkable scarcity of these compounds. Everywhere in the land and sea organisms suffer from a state of chronic nitrogen hunger, and this incessant demand for more nitrogen food than the sea or soil affords has led to a host of adaptations of modes of nutrition. It has been suggested that life first developed, or sprang into existence on the earth in the presence of accumulations of amino-acids formed after the crust of the earth first began to solidify permanently. Ammonia is still given off in the emanations from volcanoes and it may be that plutonic activity, in the presence of conditions that can never recur while the earth remains a planet, gave rise to extensive accumulations of nitrogen compounds capable of acting as the soil on which the germs of life, introduced to the earth from outer space, developed. One must suppose that such nitrogen compounds were at one time more abundant than they are now.

If they were, then it is probable that the metabolism of organisms was more wasteful than it is at the present, just as the metabolism of putrefactive and fermentative bacteria, organisms which live in a greater plenitude of food than any others, are of all the most wasteful. Such primitive organisms were probably saprophytic in habit, that is they lived by

assimilating the rich supply of food substance through
their surface. The result of their metabolism would
be the degradation of both carbon and nitrogen food
substance to the form of ammonia and carbon dioxide,
both substances on which they could no longer subsist.
It is also probable that degradation of the nitrogen
compounds would be carried on further than this
stage, and that free nitrogen would be formed, as
it is at present, by denitrifying organisms. For we
cannot but be struck by the fact that while there is
so little of nitrogen compounds in nature there is
nevertheless an enormous quantity of the element in
the atmosphere. We are tempted to regard this as
a residue formed by the degradation of compounds
of nitrogen by the agency of living organisms.

There would thus arise a scarcity of both forms
of food-stuff, carbonaceous and nitrogenous. By
developing the power of photo-synthesis plant organisms solved the question of getting sufficient carbon
food from carbon dioxide and water, under the
influence of light; and they became able to utilise
ammonia and nitrate—that is they became prototrophic in habit. Organisms developing along the
animal line continued to be saprozoic, and many
still remain so ; but this became a difficult and
precarious mode of nutrition, as it still is ; and so the
two savage methods of procuring food—by parasitism,
and by hunting and devouring other organisms—were

evolved. Even among the plants the scarcity of
nitrogen food has led to the adoption of the parasitic
habit, to their symbiotic association with animals,
and to the insectivorous method of feeding. One
must think of modes of nutrition as being just as
adaptable to changed conditions as are the structures,
habits, and life-histories of organisms.

On this view then the $79\,^0/_0$ of nitrogen which
exists in the atmosphere, and the relatively small
quantities of nitrogen compounds which are present in
the soil and sea are the results of denitrification
carried on in the past at a greater rate than at the
present time; and the tendencies of vital processes
as we know them are towards a still greater amount
of denitrification. Yet we need not be alarmed, for
living processes are essentially regulatory, and both
nature and art are combining to solve the nitrogen
question; art, in that it is now possible by the
advance of modern chemistry to combine atmospheric
nitrogen with oxygen on a large scale; and nature by
reason of the adaptability of organisms. We know
that a number of species of plants, the Legumin-
osae (peas, beans, clover, etc.), are able to take up
elementary nitrogen from the atmosphere by the
assistance of associated bacteria contained in their
root tissues. Such nitrogen-fixing bacteria have also
been discovered to be widely distributed in the sea
and on the land. They can assimilate the free gas,

and in the presence of carbon food-stuff they can synthesise it to proteid. We may then speculate further and think of a continued evolution of this mode of nutrition; that by-and-by the majority of plants and other holophytic plankton organisms may become freed from the shackles of inherited modes of nutrition; and become able to assimilate atmospheric nitrogen, just as their ancestors, who felt the pinch of carbon hunger became able to assimilate atmospheric carbon dioxide, when that gas was a residue as nitrogen is at the present. In this way we can imagine the nitrogen hunger of marine organisms becoming at last satisfied.

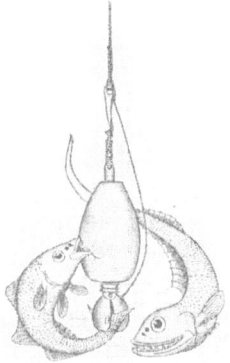

The Lucas Sounding Lead and Snapper.

APPENDICES

1. General marine biology.

Much of the matter of the present book is expanded in Conditions of Life in the Sea, Johnstone, Cambridge University Press, 1908. For greater detail as to marine zoology in general the reader may consult those volumes of the Cambridge Natural History that deal with the marine animals. The shore is easily explored by amateur naturalists and Biology of the Shore, Flattely and Walton, Sidgwick, London, 1922, is especially useful in this direction. For collecting-methods on board ship, nets, gear, etc. see Science of the Sea, G. Fowler (editor), published by Murray, London, 1912. This book is specially intended for yachtsmen. For a very good account of deep sea life in general, and for the results of modern marine biological research on the high seas, see The Depths of the Ocean, by Murray and Hjort, Macmillan, London, 1912. The above books give very many references to further literature.

2. Plankton.

There is a short general account in The Marine Plankton, Johnstone, Scott and Chadwick, the University Press of Liverpool, 1927. This book contains further references. Nordisches Plankton, Lipsius and Fischer, Kiel and Leipzig, contains detailed lists, descriptions and figures of planktonic organisms occurring in Northern Seas.

3. General oceanography.

The best small book is The Ocean by Sir John Murray, Home University Series, Williams and Norgate. See also An Introduction to Oceanography, Johnstone, 2nd edition, the University Press of Liverpool, 1928. An exhaustive book is Krümmell's Handbuch der

Ozeanographie, 2 vols., 1907, 1911, J. Engelhorn, Stuttgart. For the beginner H. R. Mill's Realm of Nature, J. Murray, London, is the best introduction. All the books in this section deal mainly with physical oceanography.

4. The ultimate food substances in the sea.

See Johnstone's Introduction to Oceanography for some general summary of the information as to the nitrates, etc. in sea water, with further references. There is much on this subject in the volumes of the Journal of the Marine Biological Association, Plymouth. (These should be seen in other connections as well.) The subject of CO_2-assimilation in the sea is very important, and the reader should study the work of Benjamin Moore (reference in The Marine Plankton). See also papers by Atkins in the Journal of the Marine Biological Association.

It should be noted, in connection with the question of the distribution of inorganic and organised food substances in the sea, that all the figures quoted in the present book are averages for large sea regions. Recent research all tends to show that both the inorganic materials, like the nitrates, and the plankton are irregularly dispersed. Near the land, nitrates, etc. may display relatively high local concentrations. The plankton is apt to occur in local 'swarms,' and fish, etc. may follow and feed upon these swarms.

5. Plant-animals.

For a general account of the problems involved, see F. Keeble's book, with the above title, in the Cambridge Manuals.

6. Denitrification in the sea.

The subject is very important. The original work is by Brandt and is published in the Reports of the Kiel Kommission (in German) in 1901 and 1902. There is an account, in English, in the American Smithsonian Report of 1901. Further very important investigations of the chemical composition of the plankton, etc. are published by Brandt in the Kiel Kommission Report of 1898. (Short summary in Johnstone's Conditions of Life in the Sea.) There is no extensive account of this very important line of work in English.

7. Habits of marine organisms, tropisms and behaviour.

The reader should see Study of Living Things by E. S. Russell, Methuen, London, 1924. For the older views regarding tropisms, see Loeb, Dynamics of Living Matter, Columbia University Biological Series, 1906. Newer conceptions are dealt with (in respect of marine animals) by Bull, Conditioned Reflexes in Fishes, Journal of the Marine Biological Association for 1928 and later. Driesch's Science and Philosophy of the Organism, Black, 1907-8, should also be read.

INDEX